应用型高等院校改革创新示范教材

土木类专门用途英语教程

主　编　宋　晶

副主编　唐敬伟　王　云　赵　凡

中国水利水电出版社
www.waterpub.com.cn
·北京·

内 容 提 要

本书为山东交通学院校本教材专门用途英语（ESP）系列教程之一，主要针对土木类专业本科生设计，作为专业英语前置课程的教材使用。本书旨在指导土木类本科生在深入学习和理解土木类主题文章的基础上，从词、句、语篇的不同层面进行听、说、读、写、译多方面的英语语言技能练习，着重培养学生的英语综合运用能力。

本书不仅包括词汇、专业术语、语法等语言知识的内容，还包括英语交际策略以及土木类职场应用型技能的练习。本书以语言能力训练为主，语言知识学习为辅，实现"翻转式"大学英语课堂教学。本书的词汇难度以大学英语四级词汇表为基准，借鉴欧洲统一语言参考框架B2等级标准，较高要求词汇可参考大学英语六级词汇表，通识性土木类专业词汇已在书中标注。本书练习的设计贴合土木语境，在充分调动学生英语学习兴趣和积极性的同时，也可为专业英语的学习奠定扎实的语言基础。

图书在版编目（CIP）数据

土木类专门用途英语教程 / 宋晶主编. -- 北京 : 中国水利水电出版社, 2021.6
应用型高等院校改革创新示范教材
ISBN 978-7-5170-9619-1

Ⅰ. ①土… Ⅱ. ①宋… Ⅲ. ①土木工程－英语－高等学校－教材 Ⅳ. ①TU

中国版本图书馆CIP数据核字(2021)第105516号

策划编辑：杜　威　　责任编辑：张玉玲　　加工编辑：庄　晨　　封面设计：梁　燕

书　　名	应用型高等院校改革创新示范教材 土木类专门用途英语教程 TUMU LEI ZHUANMEN YONGTU YINGYU JIAOCHENG
作　　者	主　编　宋　晶 副主编　唐敬伟　王　云　赵　凡
出版发行	中国水利水电出版社 （北京市海淀区玉渊潭南路1号D座　100038） 网址：www.waterpub.com.cn E-mail：mchannel@263.net（万水） sales@waterpub.com.cn 电话：（010）68367658（营销中心）、82562819（万水）
经　　售	全国各地新华书店和相关出版物销售网点
排　　版	北京万水电子信息有限公司
印　　刷	三河市航远印刷有限公司
规　　格	170mm×240mm　16开本　8.5印张　138千字
版　　次	2021年6月第1版　2021年6月第1次印刷
印　　数	0001—3000册
定　　价	25.00元

前　言

本书为山东交通学院校本教材专门用途英语（ESP）系列教程之一，主要针对土木类专业本科生设计，作为专业英语前置课程的教材使用。本书旨在指导土木类本科生在深入学习和理解土木类主题文章的基础上，从词、句、语篇的不同层面进行听、说、读、写、译多方面的英语语言技能练习，着重培养学生的英语综合运用能力。

本书不仅包括词汇、专业术语、语法等语言知识的内容，还包括英语交际策略以及土木类职场应用型技能的练习。本书以语言能力训练为主，语言知识学习为辅，实现"翻转式"大学英语课堂教学。本书的词汇难度以大学英语四级词汇表为基准，借鉴欧洲统一语言参考框架 B2 等级标准，较高要求词汇可参考大学英语六级词汇表，通识性土木类专业词汇已在书中标注。本书练习的设计贴合土木语境，在充分调动学生英语学习兴趣和积极性的同时，也可为专业英语的学习奠定扎实的语言基础。

全书共分为 6 个单元，每单元的内容探讨和学习围绕同一土木类主题展开，层层深入。每个单元的 Part II 包括 3 篇文章以及与课文相关的综合性语言练习。其中 TEXT A 作为主文章，配套练习包括以下 3 部分：

（1）Content Questions，该题为四级阅读题型段落匹配题，旨在使学生在理解课文的基础上，快速、准确地捕捉定位信息，逐步提高英语阅读能力。

（2）Vocabulary，语法结构练习帮助学生对核心词汇和术语进行理解和运用。

（3）Translation，段落翻译练习帮助学生掌握核心词汇、专业词汇，使学生对语法句法做到活学活用，提高学生对相关专业文章的翻译能力。

TEXT B 为主题扩展阅读文章，配套练习包括两部分：Content Questions，该题为判断对错题，旨在考查学生对课文篇章中重点信息的理解，帮助学生提高阅读能力；Translation，该题为汉译英翻译题，旨在帮助学生加深对原文的理解，透彻了解其中每个单词、词组的确切含义和细微差别。此部分还锻炼学生运用推理的方法，追溯作者的思路，并使用准确的词句表达出来，进一步培养学生的思考能力和逻辑推理能力。

TEXT C 文章的选用结合我国国情，注重说明与单元主题相关的我国土木领域的情况，以激发学生的思辨能力与创新思维，增强学生对于中国文化的感知力和理解力。配套练习 Translation（英译汉翻译练习）和 TEXT B 汉译英翻译练习相辅相成，帮助学生在准确掌握课文相关词汇的同时，掌握翻译技巧，提高翻译能力。

单元最后 TASKS 部分为主题学习任务，通过语言技能练习，培养学生的英语语言技能。

本书由山东交通学院宋晶担任主编，负责全书的设计、编排和书稿的审订工作，山东交通学院唐敬伟、王云、赵凡担任副主编。

本书编写思路来源于教学实践，也需要在教学实践中得到检验。书中难免存在疏漏和不当之处，望广大师生批评指正。

编者

2021 年 3 月

目　　录

Unit 1 An Introduction to Civil Engineering

Part I OVERVIEW

Civil engineering is a professional engineering discipline that deals with the design, construction, and maintenance of the physical and naturally built environment, including public works such as roads, bridges, canals, dams, airports, sewerage systems, pipelines, structural components of buildings, and railways.

Part II TEXTS

TEXT A

Development in Civil Engineering

A

The civil engineering is one of the oldest branches of engineering because of its relation to the **establishment** of built environment. It includes design, planning, and **implementation as well as** the work of developing and reconstruction and **restoration** of buildings such as bridges, **dams** and ports, also, the planning of **residential** cities and the establishing of services such as **sanitation** stations, power stations, roads and transportation network.

establishment /ɪˈstæblɪʃmənt/ *n.*
the act of starting or creating sth. that is meant to last for a long time 建立；创立；确立

implementation /ˌɪmplɪmenˈteɪʃn/ *n.*
the act of accomplishing some aim or executing some order 实施；履行

restoration /ˌrestəˈreɪʃn/ *n.*
the work of repairing and cleaning an old building, a painting, etc. so that its condition is as good as it originally was 整修；修复

dam /dæm/ *n.*
a barrier that is built across a river in order to stop the water from flowing,

used especially to make a reservoir (= a lake for storing water) or to produce electricity 水坝；拦河坝

residential /ˌrezɪˈdenʃl/ *adj.*
（of an area of a town 城市中的地区）suitable for living in; consisting of houses rather than factories or offices 适合居住的；住宅的

sanitation /ˌsænɪˈteɪʃn/ *n.*
the equipment and systems that keep places clean, especially by removing human waste 卫生设备；卫生设施体系

B

Though the expression itself was used as a scientific term in ancient Rome, it is not possible to know when did the science of civil engineering **originate**, but we can say that civil engineering has started and developed with the development of mankind through the ages and the features of engineering in **ancient** times was developed into a science to be studied till modern times.

originate /əˈrɪdʒɪneɪt/ *v.*
(formal)to happen or appear for the first time in a particular place or situation 起源；发源；发端于

ancient /ˈeɪnʃənt/ *adj.*
belonging to a period of history that is thousands of years in the past 古代的

C

For example, the **pyramids** of **Giza** pose as a model of architectural excellence and **illustrate** the development of Engineering in the **Pharaonic civilization** for several reasons, because it **consists of** 2,300,000 blocks of rock with the mass of 2 to 30 tons per block, but so far, science was unable to reach the method of how they were built. The Great Wall of China, considered one of the **Seven Wonders of the World** as it is built in less than ten years and with a length of more than 2500 kilometers.

pyramid /ˈpɪrəmɪd/ *n.*
a large building with a square or triangular base and sloping sides that meet in a point at the top. The ancient Egyptians built stone pyramids as places to bury their kings and queens（古埃及的）金字塔

illustrate /ˈɪləstreɪt/ *v.*
to make the meaning of sth. clearer by using examples, pictures, etc. （用示例、图画等）说明，解释

civilization /ˌsɪvəlaɪˈzeɪʃn/ *n.*
a state of human society that is very developed and organized 文明

D

Civil Engineering has many branches. *Construction Engineering*: It is **concerned with** the design and implementation of residential and industrial **facilities** according to different construction materials, whether metal, **concrete** or wooden.

facilities /fəˈsɪlɪtɪ/ *n.*
buildings, pieces of equipment, or services that are provided for a particular purpose 设施

concrete /ˈkɒnkriːt/*n.*
a substance used for building which is made by mixing together cement, sand, small stones, and water 混凝土

E

Transportation Engineering: It is concerned with the design and construction of roads, bridges and transportation networks.

F

Sanitary *engineering*: It is concerned with the design of **sewage** systems and water stations.

sanitary /ˈsænətrɪ/ *adj.*
connected with keeping places clean and healthy to live in, especially by removing human waste 卫生的；环境卫生的；公共卫生的

sewage /ˈsuːɪdʒ/ *n.*
used water and waste substances that are produced by human bodies, that are carried away from houses and factories through special pipes (= sewers)（下水道的）污水，污物

G

Engineering management and construction: It is interested in conducting an **inventory** of the implementation of the establishment with the lowest cost and the fastest possible time and site management.

inventory /ˈɪnvəntrɪ/ (pl. -ies) *n.*
a written list of all the objects, furniture, etc. in a particular building （建筑物里的物品、家具等的）清单；财产清单

H

Engineering of dams and water resources: It is concerned with the design of water facilities, **infrastructure** and foundations as well as **hydraulic** designs as well as **irrigation** systems.

Infrastructure /ˈɪnfrəstrʌktʃə(r)/ *n.*
the basic systems and services that are necessary for a country or an organization to run smoothly, for example buildings, transport and water and power supplies （国家或机构的）基础设施，基础建设

hydraulic /haɪˈdrɔːlɪk/ *adj.*
(of water, oil, etc. 水、油等) moved through pipes, etc. under pressure （通过水管等）液压的，水力的

irrigation /ˌɪrɪˈgeɪʃən/ *n.*
supplying dry land with water by means of ditches etc 灌溉，灌水

I

Engineering of ports and marine facilities: It is concerned with the design and implementation of seaports and **marine installations** from **docks** and **barriers,** as well as protection of beaches.

marine /məˈrɪːn/ *adj.*
connected with the sea and the creatures and plants that live there 海的；海产的；海生的

installation /ˌɪnstəˈleɪʃn/ *n.*
the act of fixing equipment or furniture in position so that it can be used 安装；设置

dock /dɒk/ *n.*
a part of a port where ships are repaired, or where goods are put onto or taken off them 船坞；船埠；码头

barrier /ˈbærɪə(r)/ *n.*
an object like a fence that prevents people from moving forward from one place to another 屏障；障碍物

J

The civil engineering has developed greatly over time. New methods have begun to appear in the design of buildings and architectural design, based, mainly, on the technological development in construction materials used and the **reliance on** environmental-friendly and stronger materials.

K

Among these **innovations** are: **Cork walls** reinforced with concrete that are used in many buildings around the world. Some of the advantages of them are saving time, which is seven times faster than construction with traditional materials, and **decreasing** the number of workers needed for the construction. This innovation makes the buildings more earthquake-**resistant** than conventional construction. It is **moisture**-resistant and fire-resistant and facilitates the change of design of the buildings.

L

A group of scientists have created the **Ultra Rope**, a rope made of **carbon fiber**, to achieve the highest rate of **rigidity**. These ropes are so strong that they can be used to **install** lifts in skyscrapers.

M

Civil engineering has reached the so-called smart buildings and buildings are characterized by several features: ①Have the ability to **detect** and react to the surrounding conditions. ②Fast response to the needs of its users. ③Create a comfortable environment

reliance /rɪˈlaɪəns/ *n.*
the state of needing sb./sth. in order to survive, be successful, etc.; the fact of being able to rely on sb./sth. 依赖；依靠；信任

innovation /ˌɪnəˈveɪʃn/ *n.*
the introduction of new things, ideas or ways of doing sth. （新事物、思想或方法的）创造；创新；改革

decrease /dɪ'krɪːsɪŋ/ *v.*
becoming less or smaller 降低，减少

resistant /rɪˈzɪstənt/ *adj.*
not affected by sth.; able to resist sth. 抵抗的；有抵抗

moisture /ˈmɔɪstʃə(r)/ *n.*
very small drops of water that are present in the air, on a surface or in a substance 潮气；水汽；水分

carbon /ˈkɑːbən/ *n.*
(symbol C) a chemical element. Carbon is found in all living things, existing in a pure state as diamond, graphite and buckminsterfullerene 碳

fiber /ˈfaɪbə/ *n.*
a slender and greatly elongated solid substance 纤维，光纤

rigidity /rɪˈdʒɪdətɪ/ *n.*
the physical property of being stiff and resisting bending 硬度，刚性

install /ɪnˈstɔːl/ *v.*
to fix equipment or furniture into position so that it can be used 安装；设置

detect /dɪˈtekt/ *v.*
to discover or notice sth., especially sth. that is not easy to see, hear, etc. 发现；查明；侦察出

for its users.

N

Dubai, Abu Dhabi and Doha are among the most cities in the Middle East to contain the most **intelligent** buildings in the region, with airports ranked first as the smartest facilities, followed by hotels and hospitals in the third place. And it is expected that by 2020, Dubai will become one of the highest and most intelligent cities in the world.

(Words: 634)

intelligent /ɪnˈtelɪdʒənt/ *adj.*
good at learning, understanding and thinking in a logical way about things; showing this ability 有才智的；悟性强的；聪明的

Useful Expressions

as well as 也；除；而且；和

consist of 由……组成；由……构成；包括

concerned with 关心；涉及；忙于；与……有关

reliance on 依靠；信赖

Proper Names

Giza/'ɡɪːzə/ 吉萨（地名）；吉萨棉；吉萨金字塔

Pharaonic 古埃及法老王的

Seven Wonders of the World 世界七大奇迹

Cork walls 软木墙

Ultra Rope 超级钢丝绳

Dubai 迪拜（阿拉伯联合酋长国之一）；迪拜港（阿拉伯联合酋长国港市）

Abu Dhabi 阿布达比酋长国（阿拉伯联合酋长国之一）；阿布扎比市（阿拉伯联合酋长国首都，另写作 Abu Zaby，Abu Zali）

Doha 多哈（卡塔尔首都）

TEXT A Exercises

1. Content Questions

Each of the following statements contains information given in one of the paragraphs in the TEXT A. Identify the paragraph from which the information is derived. You may choose a paragraph more than once. Each paragraph is marked with a letter.

1) ( ) The civil engineering is one of the oldest branches of engineering.
2) ( ) Smart buildings have the ability to detect the surrounding environment.
3) ( ) The innovation in construction makes the buildings more earthquake-resistant.
4) ( ) Civil engineering has started and developed with the development of mankind.
5) ( ) By 2020, Dubai will become one of the highest and most intelligent cities in the world.
6) ( ) Engineering of dams and water resources is concerned with the design of water facilities.
7) ( ) The Great Wall of China is considered as one of the Seven Wonders of the World.
8) ( ) Civil Engineering has many branches.

2. Vocabulary

A. Fill in the gaps with the words or phrases given in the box. Change the form when necessary.

illustrate	install	detect	consist	reliance
innovation	decrease	facility	originate	establishment
intelligent	resistant	ancient	residential	concrete

1) Last year's sales figures are __________ in Figure 2.
2) He is a highly ________child.
3) The hotel chain has recently____________ a new booking system.
4) Elderly people are not always ___________to change.

5) Population growth is___________ by 1.4% each year.

6) Heavy_______ on one client is risky when you are building up a business.

7) They produced the first vegetarian beanburger — an ________which was rapidly exported.

8) They believed _______Greece and Rome were vital sources of learning.

9) The disease__________ in Africa.

10) The ________of the regional government in 1980 did not end terrorism.

B. Fill in the gaps with the phrases in the box. Change the form when necessary.

take for granted	refer to	concerned with
consist of	reliance on	as well as

1) The lungs____________ millions of tiny air sacs.

2) We are chiefly__________ improving educational standards.

3) He plays classical music, ___________ pop and jazz.

4) About 10% of Japanese teenagers are overweight. Nutritionists say the main culprit is increasing ___________Western fast food.

5) Study this example and __________the explanation below.

3. Translation

Translate the following paragraph into English.

长期以来，建筑工程行业处于社会认知的较低水平。在一般人看来，建筑工程行业是劳动密集型的低效率行业。施工人员必须在偏远地区艰苦的工作条件下工作和生活，生活环境不稳定，远离家乡和家人。

____

____

____

TEXT B

The History of Civil Engineering

It is difficult to determine the history of **emergence** and beginning of civil engineering, however, the history of civil engineering is a mirror of the history of human beings on this earth. Man used the old shelter caves to protect themselves of weather and **harsh** environment, and used a tree trunk to cross the river, which being the **demonstration** of ancient age civil engineering.

emergence /ɪ'mɜːdʒəns/ *n.*
the gradual beginning or coming forth 出现，浮现

harsh /hɑːʃ/ *adj.*
(of weather or living conditions 天气或生活环境) very difficult and unpleasant to live in 恶劣的；艰苦的

demonstration /ˌdemənˈstreɪʃn/ *n.*
an act of giving proof or evidence for sth. 证明；证实；论证；说明

Civil Engineering has been an aspect of life since the beginnings of human existence. The earliest practices of civil engineering may have commenced between 4000 and 2000 BC in **Ancient Egypt** and **Mesopotamia** (**Ancient Iraq**) when humans started to abandon a **nomadic** existence, thus causing a need for the construction of shelter. During this time,

nomadic /nəʊˈmædɪk/ *adj.*
nomadic people travel from place to place rather than living in one place all the time 游牧的

transportation became increasingly important leading to the development of the wheel and sailing.

Until modern times there was no clear **distinction** between civil engineering and architecture, and the term engineer and architect were mainly geographical **variations** referring to the same person, often used **interchangeably**. The construction of **Pyramids** in Egypt (circa 2700-2500 BC) might be considered the first instances of large structure constructions.

distinction /dɪˈstɪŋkʃn/ *n.*
(between A and B) a clear difference or contrast especially between people or things that are similar or related 差别；区别；对比

variation /ˌveərɪˈeɪʃn/ *n.*
(in/of sth.) a change, especially in the amount or level of sth.（数量、水平等的）变化，变更，变异

interchangeably /ˌɪntəˈtʃeɪndʒəblɪ/ *adv.*
in an interchangeable manner 可交换的

Around 2550 BC, **Imhotep**, the first documented engineer, built a famous stepped pyramid for **King Djoser located at Saqqara Necropolis**. With simple tools and mathematics, he created a **monument** that stands to this day. His greatest contribution to engineering was his discovery of the art of building with shaped stones. Those who followed him carried engineering to **remarkable** heights using skill and imagination.

monument /ˈmɒnjumənt/ *n.*
(to sb./sth.) a building, column, statue, etc. built to remind people of a famous person or event 纪念碑（或馆、堂、像等）

remarkable /rɪˈmɑːkəbl/ *adj.*
~ (for sth.)~ (that...) unusual or surprising in a way that causes people to take notice 非凡的；奇异的；显著的；引人注目的

Ancient historic civil engineering constructions include the **Qanat** water management system (the oldest older than 3000 years and longer than 71 km), the **Parthenon** by **Iktinos** in **Ancient Greece** (447-438 BC), the **Appian Way** by Roman engineers (c. 312 BC), the Great Wall of China under orders from Emperor **Shih Huang Ti** (c. 220 BC) and so on. The Romans developed civil structures throughout their empire, including especially **aqueducts, insulae,** harbors, bridges, dams and roads.

aqueduct /ˈækwɪdʌkt/ *n.*
a long bridge with many arches that carries a water supply or a canal over a valley 高架渠；渡槽；桥管

insulae /ˈɪnsjʊlɪː/ *n.*
脑岛；（古罗马城市的）住屋（insula 的复数）

Machu Picchu, Peru, built at around 1450, at the height of the **Inca Empire** is considered an engineering **marvel**. It was built in the **Andes Mountains** assisted by some of history's most **ingenious** water resource engineers. The people of Machu Picchu built a mountain top city with running water, **drainage** systems, food production and stone structures so advanced that they endured for over 500 years.

marvel /ˈmɑːvl/ *n.*
a wonderful and surprising person or thing 令人惊异的人（或事）；奇迹

ingenious /ɪnˈdʒɪːnɪəs/ *adj.*
（of an object, a plan, an idea, etc. 物体、计划、思想等）very suitable for a particular purpose and resulting from clever new ideas 精巧的；新颖独特的；巧妙的

drainage /ˈdreɪnɪdʒ/ *n.*
the process by which water or liquid waste is drained from an area 排水；放水

Throughout ancient and medieval history most architectural design and construction was **carried out** by **artisans**, such as **stonemasons** and **carpenters,** rising to the role of master builder. Knowledge was retained in guilds and seldom **supplanted** by advances. Structures, roads and infrastructure that existed were **repetitive**, and increases in scale were **incremental**.

One of the earliest examples of a scientific approach to physical and mathematical problems **applicable** to civil engineering is the work of **Archimedes** in the 3rd century BC, including **Archimedes Principle,** which strengthen our understanding of **buoyancy**, and practical solutions such as Archimedes' screw. **Brahmagupta**, an Indian mathematician, used **arithmetic** in the 7th century AD, based on **Hindu-Arabic numerals,** for **excavation** computations.

In the 18th century, the term civil engineering was coined to **incorporate** all

artisan /ˌɑːtɪˈzæn/ *n.*
(formal) a person who does skilled work, making things with their hands 工匠；手艺人

stonemason /ˈstəʊnmeɪsn/ *n.*
a person whose job is cutting and preparing stone for buildings 石工；石匠

carpenter /ˈkɑːpəntə(r)/ *n.*
a person whose job is making and repairing wooden objects and structures 木工；木匠

supplant /səˈplɑːnt/ *v.*
(formal) to take the place of sb./sth. (especially sb./sth. older or less modern) 取代，替代（尤指年老者或落后于时代的事物）

repetitive /rɪˈpetətɪv/ *adj.*
saying or doing the same thing many times, so that it becomes boring 重复的；反复的

incremental /ˌɪnkrɪˈmɛntəl/ *adj.*
is used to describe something that increases in value or worth, often by a regular amount 递增的

applicable /əˈplɪkəbl/ *adj.*
~ (to sb./sth.) that can be said to be true in the case of sb./sth. 适用；合适

buoyancy /ˈbɔɪənsɪ/ *n.*
the ability that something has to float on a liquid or in the air 浮力

arithmetic /əˈrɪθmətɪk/ *n.*
the type of mathematics that deals with the adding, multiplying, etc. of numbers 算术

excavation /ˌekskəˈveɪʃn/ *n.*
the activity of digging in the ground to look for old buildings or objects that have been buried for a long time （对古物的）发掘，挖掘

incorporate /ɪnˈkɔːpəreɪt/ *v.*
~ sth. (in/into/within sth.) to include sth. so that it forms a part of sth. 将……包括在内；包含；吸收；使并入

things civilian as opposed to military engineering. The first engineering school, The National School of Bridges and Highways, France, was opened in 1747. The first self-proclaimed civil engineer was John Smeaton who constructed the **Eddystone Lighthouse.** In 1771, Smeaton and some of his colleagues formed the Smeatonian Society of Civil Engineers, a group of leaders of the profession who met informally over dinner. Though there was evidence of some technical meetings, it was little more than a social society.

In 1818, world's first engineering society, the Institution of Civil Engineers was founded in London, and in 1820 the eminent engineer **Thomas Telford** became its first president. The institution received a **Royal Charter** in 1828, formally recognizing civil engineering as a profession. Its charter defined civil engineering as: *"Civil engineering is the application of physical and scientific principles, and its history is* ***intricately*** *linked to advances in understanding of physics and*

intricately /'ɪntrəkɪtlɪ/ *adv.*
with elaboration 杂乱的

mathematics throughout history. Because civil engineering is a wide-ranging profession, including several separate ***specialized sub-disciplines****, its history is linked to knowledge of structures, material science, geography, geology, soil,* ***hydrology****, environment, mechanics and other fields."*

The first private college to teach Civil Engineering in the United States was **Norwich University** founded in 1819 by **Captain Alden Partridge**. The first degree in Civil Engineering in the United States was awarded by **Rensselaer Polytechnic Institute** in 1835. The first such degree to be awarded to a woman was granted by **Cornell University** to **Nora Stanton Blatch** in 1905.

(Word: 755)

specialized /ˈspeʃəlaɪzd/ *adj.* designed or developed for a particular purpose or area of knowledge 专用的；专业的；专门的

sub-discipline /sʌb'dɪsɪplɪn/ *n.* 学科的分支；副学科

hydrology /haɪˈdrɒlədʒɪ/ *n.* (technical 术语) the scientific study of the earth's water, especially its movement in relation to land 水文学；水文地理学

Useful Expressions

be located at 位于

carry out 执行，实施

Proper Names

Ancient Egypt 古埃及

Mesopotamia 美索不达米亚（亚洲西南部）

Ancient Iraq 古伊拉克

Pyramids 金字塔

Imhotep 英霍蒂普（公元前的埃及医生、政治家、建筑师）

King Djoser 约瑟王

Saqqara Necropolis 塞加拉墓地

Qanat 坎儿井；暗渠

Parthenon 帕特农神殿（希腊用以祭祀雅典娜女神的神庙）

Iktinos 伊克蒂诺（古希腊雅典著名建筑师）

Ancient Greece 古希腊

Appian Way 亚壁古道

Shih Huang Ti 秦始皇

Machu Picchu 马丘比丘（古城，位于秘鲁中部偏南）

Peru 秘鲁

Inca Empire 印加帝国

Andes Mountains 安第斯山脉

Archimedes 阿基米德（古希腊数学家、物理学家、发明家、学者）；阿基米德月面圆谷

Archimedes Principle 阿基米德原理

Brahmagupta(Brahmagupta) 布拉马古普塔（人名）（约公元598－655年）婆罗门笈多〈印〉天文学家

Hindu-Arabic numerals 阿拉伯数字

Eddystone Lighthouse 艾迪斯顿灯塔/涡石灯塔

Thomas Telford 托马斯·泰尔福（人名）

Royal Charter 皇家宪章（英国法律）

Norwich University 诺威奇大学

Captain Alden Partridge 奥尔登·帕特里奇队长

Rensselaer Polytechnic Institute 伦斯勒理工学院

Cornell University 康乃尔大学

Nora Stanton Blatch 诺拉·斯坦顿布拉奇

TEXT B Exercises

1. Read TEXT B and decide whether the statements are true (T) or false (F).

1) ( ) It is easy to determine the history of emergence and beginning of civil engineering.

2) ( ) Human used the old shelter caves to protect themselves of weather and harsh environment.

3) ( ) The earliest practices of civil engineering may have commenced between 4000 and 2000 BC.

4) ( ) There was no clear distinction between civil engineering and architecture.

5) ( ) The construction of Pyramids in Egypt might be considered the first instances of large structure constructions.

6) ( ) The world's first engineering society, the Institution of Civil Engineers was founded in Paris.

7) ( ) The first private college to teach Civil Engineering in the United States was Norwich University founded in 1809.

8) ( ) Archimedes Principle, strengthened our understanding of buoyancy.

2. Translation

Translate the sentences into English, using the words or phrases in brackets.

1) 可卡因上瘾者从知识丰富的人员那里得到专业化的帮助。(specialized)
2) 这种新型汽车将包含许多重大的改进。(incorporate)
3) 自然地理学是地理学的一个分支。(subdiscipline)
4) 这取决于你在挖掘过程中所做的工作,但我想我们可以做一些安排。(excavation)
5) 这是一种节约能源的巧妙方法。(ingenious)
6) 鲸,和海豚一样,已经成为造物主杰作的象征。(marvel)
7) 这项技术的成功之处在于能解决重复运动引起的问题。(repetitive)
8) 这两个酿酒区之间有明显的不同。(distinction)
9) 这是莫斯科市民对人民力量史无前例的证明。(demonstration)
10) 推动其进一步发展,即城市聚居点出现的诸多因素,是很难孤立看待的。(emergence)

TEXT C

Innovation, Policy Support Needed in Construction Industry

A flag-raising **ceremony** is held by the Civil Aviation Administration of China to mark the first **anniversary** of the Beijing Daxing International Airport in Beijing, on Sept 25, 2020.

ceremony /ˈserəmənɪ/ *n.*
(pl. -ies)a public or religious occasion that includes a series of formal or traditional actions 典礼；仪式

anniversary /ˌænɪˈvɜːsərɪ/ *n.*
(pl. -ies)a date that is an exact number of years after the date of an important or special event 周年纪念日

For a long time, the construction engineering industry was at the lower level of social cognition. To the general public, the construction engineering industry was considered labor-**intensive** yet low-**efficiency**. Construction staff has to work and live in **arduous** working conditions, suffer from an unstable living environment and through remote areas, far from home and family.

With the completion of the 13th Five-Year-Plan period, great achievements have been made in the construction industry. A

intensive /ɪnˈtensɪv/ *adj.*
involving a lot of work or activity done in a short time 短时间内集中紧张进行的；密集的

efficiency /ɪˈfɪʃnsɪ/ *n.*
the quality of doing sth. well with no waste of time or money 效率；效能；功效

arduous /ˈɑːrdʒuəs/ *adj.*
involving a lot of effort and energy, especially over a period of time 艰苦的；艰难的

large number of landmark works have been completed, including Beijing Daxing International Airport in Beijing, the Hong Kong-Zhuhai-Macao Bridge in the Guangdong-Hong Kong-Macao Greater Bay Area, as well as various expressways, railways and government foreign aid programs.

However, the development of the construction engineering industry is still **lagging** behind the pace of other emerging high-tech industries, which requires us to continue to strengthen and embrace innovation, and become a pioneer on the road toward the nation's great **rejuvenation.**

lagging /'lægɪŋ/ *n.*
used to wrap around pipes or boilers or laid in attics to prevent loss of heat 隔热材料，保温套

rejuvenation /rɪˌdʒuːvəˈneɪʃn/ *n.*
[地质][水文] 回春，返老还童；复壮，恢复活力

As for the current situation of construction projects, as a 40-year construction veteran, I think several aspects should be strengthened.

First, the construction industry is in urgent need to catch the momentum brought by the information age and fly on the wings of science and technology.

So far, the construction engineering industry is still labor-intensive – depending highly on labor with low efficiency, the participation of science and technology is not as high as other manufacturing industries since the government, society and enterprises invest less in science and technology.

We should **intensify** policy support at the national level, such as **preferential** flexible policies, tax reductions, personnel training and capital injections, to stimulate construction workers' enthusiasm to invest in scientific research fields and make positive contributions to the development and growth of construction projects.

Research institutions, colleges and universities should help to develop VR, BIM, artificial intelligence, robots, 3D printing and other technologies suitable for construction engineering. Such R&D could promote the **renewal** of the construction engineering field and **vigorously** promote the factory production of **prefabricated** buildings.

More advanced high-tech mechanical equipment suitable for construction projects could efficiently alleviate heavy physical labor.

Second, we should build up a national-level talent bank of craftsmanship to promote professionalism.

At present, most operators in the construction engineering team are migrant workers from rural areas, skilled and with decades of work experience, but many of them have only entered the engineering industry for a few years thanks to the low entry barrier.

intensify /ɪnˈtensɪfaɪ/ *v.*
to increase in degree or strength; to make sth. increase in degree or strength （使）加强，增强，加剧

preferential /ˌprefəˈrenʃl/ *adj.*
giving an advantage to a particular person or group 优先的；优惠的；优待的

renewal /rɪˈnjuːəl/ *n.*
~ (of sth.) a situation in which sth. begins again after a pause or an interruption 恢复；更新；重新开始

vigorously /ˈvɪgərəslɪ/ *adv.*
精神旺盛地，活泼地

prefabricated /ˌpriːˈfæbrɪkeɪtɪd/ *adj.*
(especially of a building 尤指建筑) made in sections that can be put together later 预制的；用预制构件组装的

Moreover, the majority of migrant workers are over 45 years old. Young people are not willing to come to the dirty and toilsome engineering industry. Such a **demographic** structure is not only inefficient, but also prone to safety and quality issues.

demographic /ˌdeməˈɡræfɪk/ *n.*
a statistic characterizing human populations (or segments of human populations) broken down by age or sex or incom etc. 人口统计数据，人口统计资料

A lack of young skilled workers leads to difficulties in proper training to cultivate skilled workers. At the national level, it is suggested training institutions and vocational and technical colleges are vigorously developed to train craftsmen. Incentives should be granted to personnel with their skills. We should encourage enterprises, industry associations and other organizations to hold skills competitions on a regular basis to select, train and encourage outstanding employees to stand out and feel the glory brought by their craftsmanship. It is necessary to form an atmosphere of **collaboration** and learning to promote progress and professionalism.

collaboration /kəˌlæbəˈreɪʃn/ *n.*
the act of working with another person or group of people to create or produce sth. 合作；协作

Third, it is necessary to ease the worries of construction engineering practitioners. The construction engineering field requires strong seasonality and high mobility. It is difficult for those in construction engineering to take good care of their families after working in the field for many years. They are the so-called local "foreigners" without a sense of belonging. If our policies are implemented better, employees, especially migrant workers, can have housing in the project sites, their children can go to school easily and they can enjoy medical insurance according to the regulations in order to solve their worries. Then, they could not only create more economic value, but also promote the development of the local area, which is conducive to the organic development of social and economic benefits.

(Word: 696)

TEXT C Exercise

Translate the following paragraph into Chinese.

A lack of young skilled workers leads to difficulties in proper training to cultivate skilled workers. At the national level, it is suggested training institutions and vocational and technical colleges are vigorously developed to train craftsmen. Incentives should be granted to personnel with their skills. We should encourage enterprises, industry associations and other organizations to hold skills competitions on a regular basis to select, train and encourage outstanding employees to stand out and feel the glory brought by their craftsmanship. It is necessary to form an atmosphere of collaboration and learning to promote progress and professionalism.

____

____

____

____

Part III TASKS

1. Viewing and Discussing

Task Video

You will watch an animation concerning civil engineering. When you finish, discuss the following questions with your partner(s). You may adopt some of the tips under each question but not limited to them.

Question 1

What is civil engineering?

Tips: Civil engineering is a professional engineering discipline that deals with design construction and maintenance of the physical and naturally built environment, including public works such as roads, brakes, canals, dams, airports, sewage system, pipelines and structural components of buildings and also railways.

Question 2

Can you list describe the evaluation in civil engineering?

Tips: It's about its evaluation from earthen dams to concrete dirty dams, from brick or stone buildings to concrete steel and time-bar buildings, from alliant camels to smooth lines channels, from building our story buildings to 828 meter tall…

Question 3

Did you get some ideas about civil engineering? Share your ideas with your partners.

2. Speech Making

Make a speech on the following topic.

Suppose, as a civil engineer you are expected to write an introduction report about civil engineering to the new staff.

You may include the following points:

1) Brief definition of Civil Engineering.
2) Branches of this industry.
3) Content of Civil Engineering.
4) Difference between Civil Engineering and other industries.
5) Your viewpoint about it.

Unit 2 Building Materials

Part I OVERVIEW

Throughout the ages, we've seen the construction industry undergo a series of building material innovations. From durable concrete used in ancient structures to the production of steel for bridges and skyscrapers, these materials shaped the way we build today. With the advancement of science and technology, there are newer cutting-edge materials being developed.

Part II TEXTS

TEXT A

Concrete Is Interesting; No Really, It Is

A

How often do you think about the impact of concrete on your life? You'd be a member of a large majority of the population if you said "never". Yet, the material that is ever-present in our world and delivers enormous economic benefits also **adds to** global warming, **contributes to** flooding, **clogs** landfills, and **releases toxic** dust. It's a challenge to make something as apparently

clog /klɒg/ *v.*
to block sth. or to become blocked （使）阻塞；堵塞

release /rɪˈliːs/ *v.*
to let sb./sth. come out of a place 释放；放出

toxic /ˈtɒksɪk/ *adj.*
containing poison; poisonous 有毒的；引起中毒的

dull as concrete interesting; so, here goes.

B

Concrete is often **erroneously referred to as cement**. Cement is an ingredient in concrete that is mixed with sand, **gravel**, and water. We don't have cement sidewalks or highways, they are concrete sidewalks and highways.

erroneously /ɪˈrəʊnɪəslɪ/ *adv.*
in a mistaken manner 错误地

cement /sɪˈment/ *n.*
a grey powder made by burning clay and lime that sets hard when it is mixed with water 水泥

gravel /ˈgrævl/ *n.*
small stones, often used to make the surface of paths and roads 沙砾；砾石；石子

C

Five thousand years ago, the Egyptians were using an early form of concrete using **gypsum** and **lime**. The Romans created a material close to modern cement to make concrete. **The Pantheon** in Rome is a concrete structure that was built 19 centuries ago and was described by **Michelangelo** as an "angelic and not human design".

gypsum /ˈdʒɪpsəm/ *n.*
a soft white mineral like chalk that is found naturally and is used in making plaster of paris 石膏

lime /laɪm/ *n.*
a white substance obtained by heating limestone, used in building materials and to help plants grow 石灰

D

In 1824, English bricklayer Joseph Aspdin invented Portland cement. He burned **clay** and limestone in his kitchen stove and

clay /kleɪ/ *n.*
a kind of earth that is soft when it is wet and hard when it is dry 黏土

crushed the result into a fine powder. It was called Portland cement because it resembled building limestone **quarried** in Portland, Dorset in southwest England. Highly **mechanized** and with many refinements this is how cement is made today.

quarry /ˈkwɒrɪ/ *v.*
stone or minerals are removed from one area by digging, drilling, or using explosives 开采；在……处开采
mechanized/ˈmekənaɪzd/ *adj.*
equipped with machinery 机械化的

E

From kitchen counter tops to skyscrapers, concrete is everywhere. More than seven billion cubic metres of the material is used every year; that's enough to provide a cubic metre of concrete to every child, woman, and man on the planet. According to the Proceedings of the **National Academy of Sciences**, the amount of concrete in the world, measured by tonnes of carbon, already **exceeds** that of all trees, bushes, and shrubs.

exceed /ɪkˈsiːd/ *v.*
greater or larger than that amount or number 超过（某数量）

F

Concrete is virtually **fireproof** and can be made waterproof. It's a bit more expensive than **asphalt** when used as a road surface, but it lasts longer and needs less maintenance. Concrete poured over steel reinforcing bars gives the material greater strength and enables it to be used for large **archways** and **domes**.

fireproof /ˈfaɪəpruːf/ *adj.*
able to resist great heat without burning or being badly damaged 防火的；耐火的
asphalt /ˈæsfælt/ *n.*
a thick black sticky substance used especially for making the surface of roads 沥青；柏油
archway /ˈɑːtʃˌweɪ/ *n.*
a passage or an entrance with an arch over it 拱门，拱道
dome /dəʊm/ *n.*
a round roof 穹顶，圆屋顶

G

We are now in a world that sounds like science fiction; in February, a 3D printed concrete pedestrian bridge was unveiled in

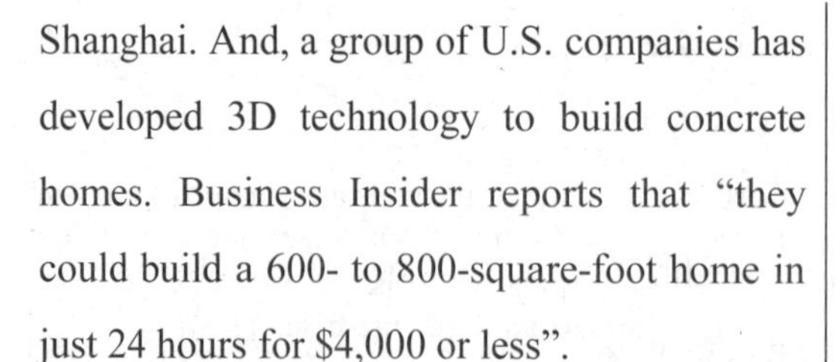

Shanghai. And, a group of U.S. companies has developed 3D technology to build concrete homes. Business Insider reports that "they could build a 600- to 800-square-foot home in just 24 hours for $4,000 or less".

H

The benefits of using concrete seem boundless; too bad there's a downside. The cement in concrete is toxic. This is the warning that Occupational Health and Safety requires on bags of dry premixed concrete: "Harmful if swallowed. Causes skin **irritation**. Causes serious eye damage. May cause allergic skin reaction. May cause cancer. May cause **respiratory** irritation. Causes damage to organs through prolonged or repeated exposure. Wash hands thoroughly after handling. Do not eat, drink or smoke when using this product. Contaminated work clothing must not be allowed out of the workplace...Wear protective gloves/protective clothing/eye protection/face protection. Use only outdoors or in a well-ventilated area. Do not breathe dust."

irritation /ˌɪrɪˈteɪʃn/ *n.*
a feeling of slight pain and discomfort in a part of your body 刺激
respiratory /rəˈspɪrətrɪ/ *adj.*
relating to breathing 呼吸的

I

According to ***The Guardian*** "concrete costs the health — and often the lives — of thousands of construction workers every year. The chief **culprit** is silica dust, which hangs in the air on building sites". The fine **particles** cause scarring in the lungs that **leads to** silicosis and a lowering of life expectancy. Silica dust has some other nasty properties that can lead to kidney disease, tuberculosis, asthma, and **chronic** obstructive pulmonary disorder.

culprit /ˈkʌlprɪt/ *n.*
a person or thing responsible for causing a problem 罪魁祸首

particle/ˈpɑːtɪkəl/ *n.*
a very small piece of sth. 颗粒；微粒

chronic /ˈkrɒnɪk/ *adj.*
lasting for a long time; difficult to cure or get rid of 长期的；慢性的；难以治愈的

J

The United Nations says that "Over the next 40 years, the world is expected to build 230 billion square metres in new construction — adding the equivalent of Paris to the planet every single week". To make way for all that construction many existing buildings are going to be **torn down**. Crushing and recycling old building materials is an option, but mostly the rubble is dumped in landfills.

K

Separating out reinforcing bars from concrete is very difficult, **not to mention**

teasing the **aggregate** away from the cement. However, it can be done cost effectively if the price of carbon is factored into using virgin materials. That only happens in a few places so most concrete **debris** just piles up in dumps.

L

The emphasis among progressive engineers and architects today is constructing buildings with the eventual **demolition** of them in mind.

(Words: 634)

aggregate /ˈæɡrɪɡət/ *n.*
sand or broken stone that is used to make concrete or for building roads, etc. （混凝土或修路等用的）骨料，集料

debris /ˈdeɪbrɪ/ *n.*
pieces of wood, metal, brick, etc. that are left after sth. has been destroyed 残骸；碎片

demolition /ˌdeməˈlɪʃn/*n.*
the act of deliberately destroying buildings 拆除；拆迁

Useful Expressions

add to 增加；加入

contribute to 有助于；促进；对某事有贡献

be referred to as 被认为是；被称为

lead to 导致，引起

tear down 拆除；拆毁

separate out 分开；析出

not to mention 更不必说

Proper Names

The Pantheon 万神殿

Michelangelo 米开朗基罗

National Academy of Sciences 美国国家科学院

The Guardian 英国《卫报》

TEXT A Exercises

1 Content Questions

Each of the following statements contains information given in one of the paragraphs in the TEXT A. Identify the paragraph from which the information is derived. You may choose a paragraph more than once. Each paragraph is marked with a letter.

1) ( ) An English bricklayer invented concrete in 19th century.

2) ( ) Concrete has been widely used and brought great economic benefits.

3) ( ) Every year thousands of construction workers lose their lives due to concrete.

4) ( ) Progressive engineers will consider the demolish of buildings when constructing them.

5) ( ) Only few old building materials are crushed or recycled.

6) ( ) The cement in concrete is poisonous and will cause many diseases.

7) ( ) Concrete is made by mixing cement with many other ingredients.

8) ( ) Egyptians and Romans created early forms of cement.

2. Vocabulary

A. Fill in the gaps with the words or phrases given in the box. Change the form when necessary.

toxic	fireproof	chronic	exceed	mechanized
quarry	debris	release	clog	irritation
particle	culprit	demolition	asphalt	clay

1) Don't ________the maximum speed of 60 km/h.

2) The new building is a ________ structure.

3) The river has been polluted with ________waste.

4) Diabetes, a_________disease, can bring serious problems.

5) Emergency teams are clearing the ________from the accident.

6) Nowadays, a good deal of farm work has been________.

7) Some cosmetics may cause ________ to sensitive skins.

8) Fast food is considered the________ of overweight.

9) The new project required the ________ of the old gate.

10) Many streets are paved with ________instead of concrete.

B. Fill in the gaps with the phrases in the box. Change the form when necessary.

add to	contribute to	be referred to as
tear down	separate out	not to mention

1) It is crucial to know how to ________ the reusable elements from the waste.

2) These new policies will________international cooperation.

3) Reading can________our knowledge of the world in the long run.

4) The Berlin Wall was________in 1989.

5) Workers in labor industries________ commonly ________ "blue-collar".

3. Translation

Translate the following paragraph into English

混凝土是城市景观的一部分，就像树木是森林的一部分。它无处不在，以至于我们很少去关注它。但是，在那个单调的灰色外表下，隐藏着一个复杂的世界。混凝土是地球上最通用、最广泛使用的建筑材料之一。它坚固耐用，防火，易于使用，而且可以制造任何形状或尺寸——从难以想象的巨大建筑到不起眼的踏脚石。

____

____

____

____

TEXT B

Is There Any Better Material Than Steel for Construction?

When it comes to construction, steel undoubtedly is the most preferred building material. Builders, **architects**, and designers have always chosen steel over other materials and with the increase in popularity of steel buildings, its demand has increased significantly. Here are some reasons that make steel a popular choice for construction projects.

architect /ˈɑːkɪtekt/ *n.*
a person who designs buildings 建筑师

1. Speedy construction

Builders and engineers want to complete construction projects on time, and steel is a time-efficient option for them. Steel components are manufactured **off-site**, and simply need to be **assembled** on site. This means that structures made from steel can be **erected** in a much lesser time, and hence steel is preferred as a building material over others.

off-site /ˈɔfˈsaɪt/ *adj.*
taking place or located away from the site 非现场的，场外的

assemble /əˈsembl/ *v.*
to fit together all the separate parts of sth. for example a piece of furniture 装配；组装

erect /ɪˈrekt/ *v.*
construct or build 建造，安装

2. Reduced cost

As steel components used in buildings are manufactured off-site, there is a limited requirement for **on site** construction, which means reduced labour costs. Also, steel does not **corrode** easily and steel buildings need minimum **maintenance**, which saves a lot in maintenance cost **as well**.

corrode /kəˈrəʊd/ *v.*
to destroy sth. slowly, especially by chemical action 腐蚀；侵蚀

maintenance/ˈmeɪntənəns/ *n.*
the act of keeping sth. in good condition by checking or repairing it regularly 维护

3. **Durability**

Steel has high **tensile** strength and has better load-carrying capacity than other construction materials. It can also **withstand** difficult external conditions like hurricanes and earthquakes which makes it an excellent construction material **in terms of** durability and safety.

durability /ˌdjʊərəˈbɪləti/ *n.*
permanence by virtue of the power to resist stress or force 耐久性

tensile /ˈtensaɪl/ *adj.*
the extent to which sth. can stretch without breaking 张力的；拉力的

withstand /wɪðˈstænd/ *v.*
to be strong enough not to be hurt or damaged by extreme conditions, the use of force, etc. 承受；抵住

4. Architectural flexibility

Another advantage of using steel in construction is that it offers scope for flexibility. For instance, if you need to expand a steel building, instead of tearing it down, you can add additional sheets of steel to achieve the desired height and length.

5. Versatility

As a construction material, steel is very **versatile**. It can be easily shaped and cut and can be used to form different building components like roofing sheets and **beams**. Its versatility makes it suitable for various kinds

versatile /ˈvɜːsətaɪl/ *adj.*
having many different uses 多用途的；多功能的

beam/biːm/ *n.*
a long thick bar of wood, metal, or concrete, especially one used to support the roof of a building 梁

of construction projects.

6. Sustainability

Structural steel is a 100% recyclable material. In fact, many steel components used in construction today are made of recycled steel. In case of concrete construction, when a building is demolished the debris becomes a part of waste and is dumped in **landfills**. But when a steel building is **dismantled**, its different components can be recycled and reused, making steel a highly eco-friendly construction material.

landfill /ˈlændfɪl/ *n.*
an area of land where large amounts of waste material are buried under the earth 废物填埋地（场）

dismantle /dɪsˈmæntl/ *v.*
to take apart a machine or structure 拆开，拆卸（机器或结构）

7. Assured quality

As steel components are manufactured in factories in a closely **monitored** and controlled environment, there's no possibility of onsite **variability**. The steel that comes out of factories undergo strict quality tests and only then they are sent to the construction site.

monitor /ˈmɒnɪtə/ *v.*
to check sth. over a period of time in order to see how it develops 监视；检查

variability /ˌveərɪəˈbɪlətɪ/ *n.*
the fact of sth. being likely to vary 可变性；易变性

Steel, **without a doubt**, offers a host of benefits and is certainly sturdier and more durable than other construction materials. It can make your construction withstand the test of time without negatively impacting the environment.

(Words:450)

Useful Expressions

when it comes to 当提到，就……而言

on site 现场，原地

as well 也，同样地

in terms of 在……方面，就……而言

without a doubt 无疑地

TEXT B Exercises

1. Read TEXT B and decide whether the statements are true (T) or false (F).

1) ( ) To complete construction projects with steel is time-consuming.

2) ( ) It will cost a lot of money to keep steel buildings in good condition.

3) ( ) Steel structure can resist extreme weather conditions.

4) ( ) If you want to expand a steel building, you must tear it down.

5) ( ) Steel can be easily shaped into building components like beams.

6) ( ) Few steel components used in construction today are made of recycled steel.

7) ( ) The steel that comes out of factories will be sent to the construction site directly.

8) ( ) Steel buildings have no bad influence on environment.

2. Translation

Translate the sentences into English, using the words or phrases in brackets.

1) 我们的大部分业务都是通过非现场会议完成的。（off-site）

2) 这个书架容易组装。（assemble）

3) 这座纪念碑是为纪念牺牲的战士们而建立的。（erect）

4) 这种液体会腐蚀铁。（corrode）

5) 精心保养可延长汽车的寿命。（maintenance）

6) 科技的进步提高了产品的耐用性。（durability）

7) 公众人物需要承受比普通人更多的压力。（withstand）

8) 他是一位多才多艺的建筑师。（versatile）

9) 提到成功，勤奋是最重要的。（when it comes to）

10) 就薪水而言，这份工作很棒。（in terms of）

TEXT C

Chinese Wooden Architecture: Prevalence & Decline

Wooden architecture was the mainstay in traditional Chinese building. Wood was preferred for most traditional architecture, from the halls of the Forbidden City to common houses. But why?

Wood's Advantages — 4 Basic Reasons for Wooden Architecture in China

Here are the four main reasons why wooden buildings **prevailed** in China, up until the modern era.

prevail /prɪˈveɪl/ *v.*
to exist or be very common at a particular time or in a particular place 普遍存在；盛行；流行

1. Wood was an abundant resource in early China.

The first reason that Chinese have a preference for wooden structures is tied up with the abundance of forests in Chinese civilization's birthplaces — the Yellow River and Yangtze River valleys. **Archaeological** evidence of wooden stilt houses in these was are as have been dated as far back as 7,000 years ago (the Hemudu Culture).

archaeological /ˌɑːkiəˈlɒdʒɪkl/ *adj.*
related to or dealing with or devoted to archaeology 考古学的

2. Ancient Chinese philosophy states that wood is lucky.

Wood remained the most popular building material even after quarrying and brickmaking developed, due to the Five Elements Theory used in fengshui (geomancy), which has dictated many aspects of life since the Spring and Autumn Period (770–476 BC). As wood is the element that represents spring and life, it has the best **auspicious connotations** for buildings. So fengshui believers have felt **compelled** to build their houses etc. out of wood.

3. Wood was easy to produce.

With a relatively short growing period for most trees used, demand didn't **outstrip** supply as China's population grew. Wood remained the ideal building material: easy to obtain, process, and replenish. Some dynasties decreed that each family should plant some trees to ensure the ready supply of China's construction material of choice.

4. Wood was easy to work with.

The emperor's throne in the Forbidden City Traditional Chinese wooden architecture became more complicated and ornate in the imperial era (221 BC – 1912 AD). As Chinese culture developed, its architecture became more complicated and ornate. Coupled with a huge population growth, more people wanted more, and only wood could keep up with demands. The good workability of wood made traditional Chinese architecture's building

auspicious /ɔːˈspɪʃəs/ *adj*.
showing signs that sth. is likely to be successful in the future 吉利的；吉祥的
connotation /ˌkɒnəˈteɪʃn/ *n.*
an idea suggested by a word in addition to its main meaning 含义；隐含意义
compel /kəmˈpel/ *v.*
to force sb. to do sth.; to make sth. necessary 强迫；迫使；使必须
outstrip/ˌaʊtˈstrɪp/ *v.*
to become larger, more important, etc. than sb./sth. 比……大（或重要等）；超过；胜过

speed much faster than other civilizations' structures of stone and mortar. Decorations were also easier to form.

Chinese buildings were usually finished in several years, while other civilizations' needed decades. However, the buildings of the Romans and other ancient civilizations generally lasted much longer.

Wood's Disadvantages — 4 Reasons for Wooden Architecture's Decline in China

1. Wood's Perishability—High Maintenance

Exposure to the elements — sun, wind, and rain — not to mention insect attack and other abrasive forces, meant that wooden structures quickly **deteriorated**. Even with the invention of paint and other preservative measures, wooden structures needed frequent repair and replacement work. High-initial- outlay but low-maintenance materials, like bricks and stone (and now concrete) became increasingly attractive as the cost of labor increased in China.

deteriorate /dɪˈtɪərɪəreɪt/ *v.* to become worse 变坏；恶化；退化

2. Wood's Flammabilit— Building Safety

Another shortcoming of wooden constructions is wood's vulnerability to fire. According to records from the Ming Dynasty to the Qing Dynasty (1368–1912), there were over 50 major fires in China's cities! The fire risk has meant that wooden structures are not so sought after anymore. The values of safety and security have outstripped those of easy construction, aesthetics, and even philosophical beliefs in China.

3. Pressure on Forest Resources— Not Enough Wood

In China's modern era of rapid development and dense population, land has been increasingly used for farming, transport, industry, and housing, and forest resources have been severely depleted. Nowadays, protecting forest resources has been made public policy in China to meet the demands of sustainable development — to stop **deforestation** and provide for China's much reduced demand for wood in buildings.

deforestation /ˌdiːˌfɔːrɪˈsteɪʃn/ *n.*
the act of cutting down or burning the trees in an area 毁林；滥伐森林；烧林

4. Wood's Lack of Strength — No Good for Tall Buildings

A final reason for the decline in one- or two-story wooden buildings is the pressure to build taller with China's 1.4 billion people. Multi-story buildings of brick and reinforced concrete seem the only viable solution to comfortably house the masses. Wood simply is not strong enough for modern construction.

(Words: 644)

TEXT C Exercise

Translate the following paragraph into Chinese.

Ancient Chinese philosophy states that wood is lucky. Wood remained the most popular building material even after quarrying and brickmaking developed, due to the Five Elements Theory used in fengshui (geomancy), which has dictated many aspects of life since the Spring and Autumn Period (770–476 BC). As wood is the element that represents spring and life, it has the best auspicious connotations for buildings. So fengshui believers have felt compelled to build their houses etc. out of wood.

____

____

____

____

____

____

Part III TASKS

1. Viewing and Discussing

You will watch a video concerning new building materials. When you finish, discuss the following questions with your partner(s).

Task Video

Question 1

What's special about this new concrete?

Question 2

Can you describe what kinetic paving material is?

Question 3

What do you think of the 4D printing?

2. Writing

You are going to write an essay on the topic *New Trends of Building Materials.* You should write the essay according to the following structures:

1) The traditional building materials;
2) Introduce some new building materials;
3) Summarize the new trends of building materials.

When you finish, exchange your writing with a partner and evaluate each other's writing according to the following standards:

1) You must write an essay of at least 120 words.
2) You must use transitional words and phrases to guide readers through your analysis.
3) Try to use as many as possible of new words and expressions.

Unit 3 Construction Project Management

Part I OVERVIEW

Construction management is a long and extremely demanding process. It's the foundation for every building project and the key to its success. The construction manager must sharply control and monitor the progress of a project in terms of quality, cost and time. Besides, the sustainability should also be given enough attention in the construction process.

Part II TEXTS

TEXT A

Construction Project Management

A

According to the Project Management Institute (PMI), project management is the "art of directing and **coordinating** human and material resources throughout the life of a project by using modern management techniques to achieve predetermined objectives of scope, cost, time, quality, and participating objectives".

coordinate /kəʊˈɔːdɪneɪt/ *v.*
to organize the different parts of an activity and the people involved in it so that it works well 使协调；使相配合

B

Construction project management typically includes complicated tasks that can **shift**

shift /ʃɪft/ *v.*
move or change slightly 改变；变换

wildly, **depending on** the work **at hand**, and it requires strong skills in communication, deep knowledge of the building process, and the ability to problem-solve. Construction project management is a complex field, requiring knowledge in many different areas like finance, **mediation**, law, business, and more.

mediation /ˌmiːdiˈeɪʃn/ *n.*
negotiation to resolve differences 调解

C

As long as there have been complex building projects, there have been project managers. For centuries, however, the person **overseeing** the construction of a complex building was often the architect, which is thought to be the case in ancient structures like the **Great Pyramids** of **Egypt** and the **aqueducts** of **Rome**.

oversee /ˌəʊvəˈsiː/ *v.*
to watch sb./sth. and make sure that a job or an activity is done correctly 监督；监视

aqueduct /ˈækwɪdʌkt/ *n.*
a structure for carrying water, usually one built like a bridge across a valley or low ground 渡槽；高架渠

D

Into the **Renaissance**, individual architects began to be known for their designs, like Sir Christopher Wren of England. Wren designed and built buildings in the late 17th and early 18th centuries, including the masterpiece **St. Paul's Cathedral**, that help give London its rich **countenance**. Wren had a breadth of knowledge that would **foreshadow** the types of skills needed on a complicated construction project, with **expertise** in advanced mathematics and physics, as well as in design. He was on his building sites every

countenance /ˈkaʊntənəns/ *n.*
a person's face or their expression 面容；面部表情

foreshadow /fɔːˈʃædəʊ/ *v.*
to be a sign of sth. that will happen in the future 预示；是……的预兆

expertise /ˌekspɜːˈtiːz/ *n.*
expert knowledge or skill in a particular subject, activity or job 专门知识；专门技能

day overseeing every **phase** of the works.

phase /feɪz/ *n.*
a stage in a process of change or development 阶段；时期

E

The rules of project management began to **take shape** across corporate America around the time of **World War II**, and by the 1950s, they were guiding civil construction projects. This meant that the phases and **tenets** of managing a construction engineering project were now being applied to a variety of corporate projects.

tenet /ˈtenɪt/ *n.*
the principles or beliefs 原则；信条

F

More and more details of managing a construction project can be done digitally, and that trend is expected to grow. Mobile-friendly technology and software are set to **play a major role in** the field, as a younger workforce is more comfortable with the technology, and it will allow the work to be managed and **tracked** from anywhere.

track /træk/ *v.*
to follow the progress or development of sb./sth. 跟踪（进展情况）

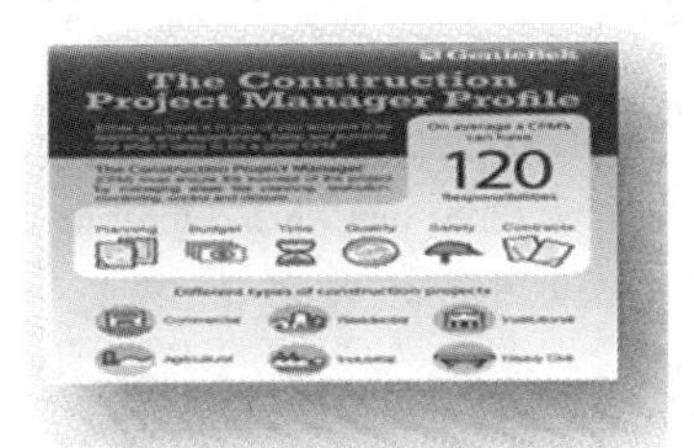

G

Although the stages of a construction project are different than that of traditional project management, they follow a similar pattern. All construction project managers should **familiarize** themselves **with** the five phases of project management, as developed by the Project Management Institute.

H

Initiation. Before the project starts, a project manager must develop and evaluate the business case to determine if the project is feasible and worth undertaking. **Stakeholders** may be asked to do their due diligence and to conduct feasibility testing, if needed. When all parties agree to **proceed** with the project, the project manager writes a project **charter** or project initiation document (PID), which includes both the business needs and the business case.

initiation /ɪˌnɪʃɪˈeɪʃn/ *n.*
(formal) the act of starting sth. 开始；创始；发起

stakeholder /ˈsteɪkhəʊldə/ *n.*
a person or company that is involved in a particular organization, project, system, etc., especially because they have invested money in it 股东；利益相关者

proceed /prəˈsiːd/ *v.*
to continue doing sth. that has already been started 继续做（或从事、进行）

charter /ˈtʃɑːtə(r)/ *n.*
a formal document describing the rights, aims, or principles of an organization or group of people 章程

I

Planning. Next, the project team develops a road map for all involved. This includes the project management plan (PMP), a formal, approved document created by the project manager to guide execution and control, as well as set **baselines** for scope, cost, and schedule.

baseline /ˈbeɪslaɪn/ *n.*
a line or measurement that is used as a starting point when comparing facts 基准

J

Execution. Now the work begins. Typically, all parties hold a **kickoff** meeting, then the project team begins the crucial work of **assigning** resources, **implementing** project

execution /ˌeksɪˈkjuːʃn/ *n.*
the act of doing a piece of work, performing a duty, or putting a plan into action 实行；执行；实施

kickoff /ˈkɪkˌɔːf/ *n.*
the time at which something is supposed to begin 开始

management plans, setting up tracking systems, completing tasks, updating the project schedule, and if necessary, **modifying** the project plan.

K

Performance and Monitoring. The monitoring phase often happens concurrently with the execution phase. This phase is necessary to measure progress and performance and to ensure that items are **in line with** the overall project management plan.

L

Closure. This final phase marks the project's completion. To mark the conclusion, project managers may hold a **post-mortem** meeting to discuss what parts of the project did and didn't meet objectives. The project team then creates a punch list of any **lingering** tasks, performs a final budget, and issues a project report.

(Words: 650)

assign /ə'saɪn/ *v.*
the act of distributing something to designated places or persons 分派，分配

implement /'ɪmplɪm(ə)nt/ *v.*
to make sth. that has been officially decided start to happen or be used 执行；实施

modify /'mɒdɪfaɪ/ *v.*
change it slightly, usually in order to improve it 修改

post-mortem /ˌpəʊst 'mɔːtəm/ *adj.*
a discussion or an examination of an event after it has happened 事后反思（或剖析）

lingering /'lɪŋgərɪŋ/ *adj.*
slow to end or disappear 拖延的；迟迟不去的

Useful Expressions

depend on 取决于，依赖

at hand 在手边，即将到来

take shape 形成，成型

play a major role in 发挥重要作用

familiarize …with 使熟悉

in line with 与……一致，符合

Proper Names

Great Pyramids 大金字塔

Egypt 埃及

Rome 罗马

Renaissance 文艺复兴

St. Paul's Cathedral 圣保罗大教堂

World War II 第二次世界大战

TEXT A Exercises

1. Content Questions

Each of the following statements contains information given in one of the paragraphs in the TEXT A. Identify the paragraph from which the information is derived. You may choose a paragraph more than once. Each paragraph is marked with a letter.

1) ( ) In monitoring phase, progress and performance of the project will be measured.

2) ( ) For centuries, it's the architect who supervises the construction of a complex building.

3) ( ) The rules of project management was applied to civil construction projects in 1950s.

4) ( ) Project managers use modern management techniques to achieve objectives.

5) ( ) The punch list records the lingering work of the project.

6) ( ) In the kickoff meeting, the project team starts some important work.

7) ( ) Sometimes, stakeholders may be asked to check the feasibility of the project.

8) ( ) Construction project management requires the ability to handle various problems.

2. Vocabulary

A. Fill in the gaps with the words or phrases given in the box. Change the form when necessary.

coordinate	shift	mediation	oversee	expertise

phase	track	foreshadow	stakeholder	proceed
baseline	execution	assign	lingering	modify

1) To grow the business, he needs to develop management________and innovation across his team.
2) The dispute has been settled through ________.
3) You need to ________ arms, legs, and breathing for the front crawl.
4) The research tried to ________ the careers of 1000 graduates.
5) We're not sure whether we should ________with the investigation.
6) The teacher would ________a topic to students to write an essay from time to time.
7) The policy was good, but the ________was poor.
8) You'd better ________the file name to distinguish it from other files.
9) If you want to drive faster, you should ________ gears.
10) The ________ shadow of the global financial crisis makes it harder to accept extravagant indulgences.

B. Fill in the gaps with the phrases in the box. Change the form when necessary.

depend on	at hand
take shape	play a major role
familiarize…with	in line with

1) Please keep an English dictionary________when you take this examination.
2) You need time to ________yourself ________our procedures.
3) The United Nations ________in the world's peacekeeping.
4) We need to know whether our thoughts are ________those of other people.
5) Owing to our joint efforts, our plan is beginning to________.

3. Translation

Translate the following paragraph into English.

据一份报告所说，一名项目施工经理有多达 120 项不同的职责。简单地说，施工经理是负责工程按照现有计划顺利进行的人。施工经理的主要任务是管理他们的项目，确保其在商定的预算和时间内完成。此外，他们还应确保整个项目符合既定的建筑计划、规范和其他规定。

____

____

____

____

____

____

TEXT B

Techniques for Sustainable Building Construction

A quieter part of the **sustainability** story is the evolution in construction techniques and materials acquisition that can reduce waste, energy and various inefficiencies at building sites.

sustainability /səˌsteɪnəˈbɪlətɪ/ *n.*
the property of being sustainable 可持续性

For **contractors**, a strategy for saving time and materials can lead to higher profitability and the good feeling of not creating unnecessary waste. Here's a look at five techniques that are having the greatest impact on sustainable building construction.

contractor /kənˈtræktə(r)/ *n.*
a person or company that has a contract to do work or provide goods or services for another company 承包人；承包商；承包公司

1. **Prefabricating** Materials in Controlled

prefabricate /priːˈfæbrɪkeɪt/ *v.*
to manufacture sections of (a building), especially in a factory, so that they can be easily transported to and rapidly assembled on a building site （尤指在工厂中）预先制造（建筑物的部件）

Environments.

"Constructing as much of a structure in a controlled environment as possible has improved the quality of buildings and resulted in less trash," says Spencer Finseth, principal of a construction company. "Being able to cut materials precisely decreases waste and creates buildings that are strong enough to allow contractors to use wood framing as high as five stories," he says.

Mechanical contractors use Building Information Management (BIM) systems to cut sheet metal for **duct** work in a controlled environment instead of outside to avoid the shape-changing problems caused by cold or hot weather, according to Mike Smoczyk, a director of professional development. That same duct work is delivered to a project "wrapped and **sealed** tightly and kept out of the elements to avoid damage," he says. He estimates that prefabrication probably accounts for 15% of any project and likely more for hotels.

2. Construction Waste Management

Reducing waste is becoming more achievable for contractors as **haulers** have grown more **sophisticated** in recent years. Where jobsites once had trash bins for different types of waste, they now need just

duct/dʌkt/ *n.*
a pipe, tube, or channel which carries a liquid or gas 输送管道

seal /siːl/ *v.*
to close a container tightly or fill a crack, etc., especially so that air, liquid, etc. cannot get in or out 密封（容器）

hauler /ˈhɔːlər/ *n.*
an automotive vehicle for hauling goods or material 运输车

sophisticated /səˈfɪstɪkeɪtɪd/ *adj.*
(of a machine, system, etc. 机器、体系等) clever and complicated in the way that it works or is presented 复杂巧妙的；先进的；精密的

one, in many cases, because haulers use pickers to separate materials. “Through haulers, we can achieve 75% landfill avoidance through their process and we don’t need to separate materials to do it,” says Dale Forsberg, president of a construction company. “On a couple of sites, we’ve hit 95%.”

For inner city projects with small footprints, having haulers handle materials in a single container **makes all the difference** because space is **at a premium**, Forsberg says. Some materials are recyclable on site — **in particular**, concrete that can be crushed and used for foundations or as aggregate beneath parking lots.

3. Managing the Site for Improved Environment

Stormwater pollution prevention has become a “big deal” to **municipalities** and the state and federal government, says Smoczyk from a construction company. “Municipalities do not want a [construction] development that dumps a bunch of bad water into the **storm sewer system** and overflows it. **Runoff** is now contained by **silt** fencing surrounding an area. A number of “best practice” approaches can be used to treat water on site and avoid having it flow into the local sewer system,” he says.

Forsberg says worker safety has led to

municipality /mjuːˌnɪsɪˈpælɪtɪ/ *n.*
a town, city or district with its own local government; the group of officials who govern it 自治市；（市下的）自治区；市（或区）政当局

runoff /ˈrʌnɒf/ *n.*
the occurrence of surplus liquid (as water) exceeding the limit or capacity 径流

silt /sɪlt/ *n.*
fine sand, soil, or mud which is carried along by a river 泥沙

restrictions and the institution of simple ways to reduce pollution. There's no smoking on the site, for example. When workers enter a building, they travel over "walk-off mats" that remove dirt, lead and other potentially dangerous chemicals from their shoes. Contractors also bring recycling containers for food to decrease organic waste.

4. Lean Manufacturing to Reduce Energy

Brenteson, an engineer, says his company encourages rethinking construction approaches through lean thinking. "It's finding the wasteful activities we're doing and eliminating them," he explains. "It saved a substantial amount of time and manpower and that's important when talking about waste and sustainability."

LEED doesn't give contractors points for lean construction techniques, but many contractors use them anyway. Ted Beckman says his company sits down with **foremen** from various subcontractors to share schedules so "everyone knows what they're responsible for".

The materials are delivered "just in time" to avoid having **rebar** and other materials sitting outside well before **installation**. The just-in-time system brings supplies on or around the day they are needed, Beckman

foreman /'fɔːmən/ *n.*
a person, especially a man, in charge of a group of workers（尤指男性）工头

rebar /'rɪːbɑː/ *n.*
a steel rod with ridges for use in reinforced concreter 钢筋

installation/ˌɪnstəˈleɪʃn/ *n.*
the act of fixing equipment or furniture in position so that it can be used 安装；设置

says. “It saves time, eliminates theft on the jobsite, eliminates damage, eliminates wasted time moving things,” he adds. “Those are lean practices but they are sustainable things, too, in a sense.”

5. Material Selection

Architects and clients seeking LEED can achieve many points by selecting materials manufactured from recycled products and from local sources. “The materials can be anything, from renewable products such as bamboo for floors”, to wood from vendors approved by the Forest Stewardship Council. LEED points are also available for installing water-saving dual-flush toilets and low-flow **faucets** and other features,” says Smoczyk. “Water reduction has become a major issue, even in the Land of 10,000 Lakes,” he notes.

As buildings become greener, so do construction sites. Off-site fabrication, improved on-site maintenance, lean practices, landfill avoidance and green materials acquisition have begun to fundamentally, **albeit** slowly, transform the way buildings are constructed today.

(Words:779)

faucet/ˈfɔːsɪt//*n.*
a device that controls the flow of water from a pipe 龙头；旋塞

albeit/ˌɔːlˈbiːɪt/ *conj.*
(formal) although 尽管；虽然

Useful Expressions

makes all the difference 大不相同；关系重大

at a premium 稀缺的

in particular 尤其，特别

Proper Names

storm sewer system 雨水管网系统

LEED 能源与环境设计先锋（绿色建筑评价体系）

TEXT B Exercises

1. Read TEXT B and decide whether the statements are true (T) or false (F).

1) ( ) The evolution in construction techniques can improve the profitability.

2) ( ) Constructing a structure in a controlled environment will increase waste.

3) ( ) The weather change might influence the metal cutting.

4) ( ) Different from the past, nowadays there are different types of trash bins in the construction site.

5) ( ) The using of haulers can help to reduce landfill.

6) ( ) Some dangerous chemicals from the workers' shoes will be eliminated when they step into the building site.

7) ( ) Lean manufacturing cannot be sustainable.

8) ( ) Architects and clients seeking LEED only select materials from recycled products.

2. Translation

Translate the sentences into English, using the words or phrases in brackets.

1) 在选择购买产品时，更多消费者开始关注可持续性。（sustainability）

2) 合同规定了雇主与承包人应该承担的风险。（contractor）

3) 这些蔬菜被密封在塑料袋里。（seal）

4) 没人可以操作如此复杂的机器。（sophisticated）

5) 我们这个市包含 10 个区。（municipality）

6) 新设备的安装需要几天时间。（prevalent）

7) 预先制造建筑配件可以节省时间。（prefabricate）

8) 有时一个简单的改变可以改变一切。（make all the difference）

9) 对于上班的父母来说，空余时间是非常稀缺的。（at a premium）

10) 发达国家尤其应该对环境问题承担责任。（in particular）

TEXT C

China Speed：Hospital Construction Shows Nation's Building Prowess

Laboring day and night, thousands of workers and hundreds of machines constructed Wuhan's Huoshenshan Hospital — a 1,000-bed emergency field hospital designed to tackle the novel coronavirus — in just nine days.

The government of Wuhan, in Central China's Hubei Province, which is the **epicenter** of the coronavirus outbreak, decided to build the Huoshenshan Hospital and the Leishenshan Hospital to relieve pressure on local hospitals and ensure the best treatment of coronavirus patients.

From design to **excavation**, from setting up internet **coverage** to opening a 24-hour smart convenience store, the construction of the hospitals was made possible by the contributions of professionals from disparate sectors who together used their industrial and technological expertise to pull off the **mammoth** urgent task. It put China's massive

epicenter /ˈepɪsentər/ *n.*
the point on the Earth's surface directly above the focus of an earthquake 震中

excavation/ˌekskəˈveɪʃn/ *n.*
the act of digging, especially with a machine 挖掘；开凿；挖土

coverage /ˈkʌvərɪdʒ/ *n.*
the reporting of news and sport 新闻报道

mammoth /ˈmæməθ/ *adj.*
extremely large 极其巨大的；庞大的

manufacturing, construction, design and organizational abilities and knowhow on full display.

On the first day of construction, which was January 24 and this year's Chinese New Year's eve, a team of 95 excavators, 33 bulldozers, five heavy rollers, 160 dump trucks, 160 managers, 2,240 workers was assembled. That night at the site, 50,000 square meters — the size of seven soccer fields, — was leveled.

Busy work

In livestreams from the construction site, provided by China Central Television, millions of online viewers made up nicknames for the machines they could see working through these days.

Concrete pump trucks were **dubbed** "Dahong", or Big Red. The excavators were named "Xiaohuang", or Little Yellow. During construction 106 pieces of heavy equipment worked around the clock building the foundation for the hospital.

dub /dʌb/ *v.*
to give sb./sth. a particular name, often in a humorous or critical way 把……戏称为；给……起绰号

"Our drivers who had returned to their hometowns for the Spring Festival holiday returned to work in Wuhan after receiving phone calls from the company," said Chen Jinsong, an employee with a Hubei machinery company. "Some of the work needed two,

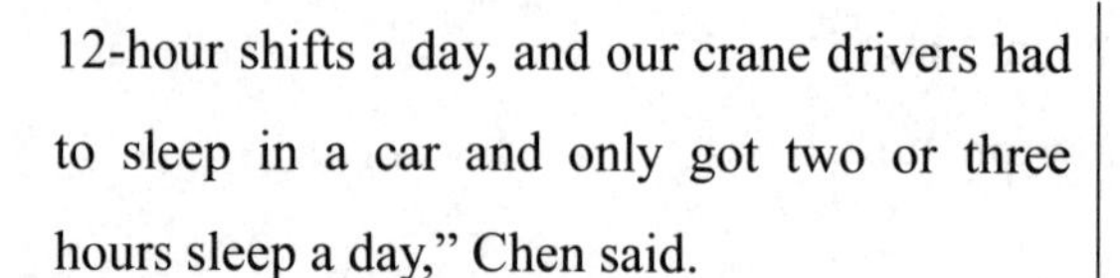

12-hour shifts a day, and our crane drivers had to sleep in a car and only got two or three hours sleep a day," Chen said.

"Due to urgency there were some serious construction challenges such as the prefab buildings ordered from various suppliers had different design standards," said Huang Tian, a project technician. "A difference of a couple of centimeters can add up to several meters, when hundreds of the prefab buildings are spliced together," Huang said.

In order to solve this problem, more than 780 containers had to be "precisely positioned", which Huang and her colleagues managed complete after four **consecutive** days and nights.

consecutive/kənˈsekjətɪv/ *adj*. following one after another in a series, without interruption 连续不断的

Electricity guarantee

On early Saturday morning, Huoshenshan Hospital's electric power was turned on. More than 200 construction personnel from State Grid Corp's Wuhan branch spent about a week, often in inclement weather, completing the electric wiring of every room in the hospital.

"To ensure the safe use of electricity in Leishenshan Hospital, the power company assigned safety managers and set up fences and tape warning other workers in potentially dangers areas, creating safe construction site."

Hu Hao, general manager of the local power supply firm in the Caidian district under the State Grid said.

"The fast construction of Huoshenshan and Leishenshan hospitals impressed the nation, showing off China's strong infrastructure muscle, organizational and coordination skills, resources and sound supply chain guarantees," Yang Xiaowen, a senior regional manager responsible for Wuhan with Schneider Electric, told the Global Times.

"All workers on site wear masks and conduct construction work in an orderly fashion," Yang said. Before the Huoshenshan Hospital began to receive patients on Monday, the electrical equipment had been set up in the hospital.

"Generally, it takes about 1-2 weeks from winning the bid to delivering core components of a project, but all the core components needed in the construction of Huoshenshan Hospital had to be delivered within two days. And construction materials of Leishenshan Hospital need to be delivered within three days." Yang said, noting that the firm conducted "unconventional operations" to secure the supply of electric parts.

"Everyone worked very hard... I believe

that no matter wherever there is difficulty in China, every Chinese citizen has an obligation and is willing to contribute." Liu Tao, an employee with SANY said.

(Words: 644)

TEXT C Exercise

Translate the following paragraph into Chinese.

From design to excavation, from setting up internet coverage to opening a 24-hour smart convenience store, the construction of the hospitals was made possible by the contributions of professionals from disparate sectors who together used their industrial and technological expertise to pull off the mammoth urgent task. It put China's massive manufacturing, construction, design and organizational abilities and knowhow on full display.

____

____

____

____

____

Part III TASKS

1. Discussing

Suppose you are a construction contractor, what measures would you take to make the construction process sustainable or green? Discuss this question with your partners and do a presentation.

2. Writing

You are going to write an essay on the topic *Qualities That A Construction Project Manager Should Have.* You should write the essay according to the following structures:

1) Briefly describe the content of construction management.
2) List some crucial qualities that a construction project manager should have.
3) Conclude your points.

When you finish, exchange your writing with a partner and evaluate each other's writing according to the following standards:

1) You must write an essay of at least 120 words.
2) You must use transitional words and phrases to guide readers through your analysis.
3) Try to use as many as possible of new words and expressions.

Unit 4 Architectural Design

Part I OVERVIEW

Architectural design is a concept that focuses on components or elements of a structure. An architect is generally the one in charge of the architectural design. They work with space and elements to create a coherent and functional structure. An architectural design focuses on the aesthetic and function of the structure and it aims to create a space that flows with its surroundings.

Part II TEXTS

TEXT A

Smart Architectural Design for a Green Future

A

There was a time when the **annual** World's Fair was the planet's primary showcase for "the buildings of tomorrow". People from around the globe gathered in futuristic national **pavilions** surprised at what the future of construction might bring.

B

Today, concern for the environment balances that **enthusiasm** for progress. Where

annual /ˈænjʊəl/ *adj.*
events happen once every year 每年一次的

pavilion/pəˈvɪljən/ *n.*
temporary building, especially a large tent used to display items at an exhibition（展览会陈列物品用的）临时建筑物，（尤指）大帐篷

enthusiasm /ɪnˈθjuːzɪˌæzəm/ *n.*
great eagerness to be involved in a particular activity that you like and enjoy or that you think is important 热情

they once aimed for **Space Needles** and flying-saucer-like **observation towers**, construction industry leaders now focus on **innovations** that make buildings greener and more responsive to the needs of occupants—thinking smaller, you might say, in order to see the bigger picture.

innovation /ˌɪnəˈveɪʃən/ *n.*
the introduction of new ideas, methods, or things 创新；革新

C

Industry research by Kingspan, a global leader in high-performance **insulation** and building envelopes, **confirms** the need for this **evolution** in thinking. It shows that the construction and operation of buildings **account for** 36% of global energy use. It also shows that construction accounts for 39% of energy-related CO2 **emissions** and approximately 30% of the waste sent to global landfills.

insulation /ˌɪnsjʊˈleɪʃən/ *n.*
material used to insulate something, especially a building 隔热材料

confirm /kənˈfɜːm/ *v.*
to establish or strengthen as with new evidence or facts 证实

evolution /ˌiːvəˈluːʃn/ *n.*
a process of gradual development in a particular situation or thing over a period of time 演化；发展

emission /ɪˈmɪʃən/ *n.*
the release of something such as gas or radiation into the atmosphere 排放

D

An Environment for the Mind

This research informed the design and construction of IKON, Kingspan's global innovation center. Opened in 2019 amid the lakes and rivers of County Cavan, north of

Dublin, IKON **represents** an investment of 10 million euros in Kingspan's **commitment** to a greener future. It functions as a "living experiment", providing engineers with the perfect real-world setting to measure the energy-saving properties of new materials.

represent /ˌreprɪˈzent/ *v.*
to act on behalf of a person, group, or place 代表

commitment /kəˈmɪtmənt/ *n.*
a promise to do something or to behave in a particular way 承诺

E

"The original idea for IKON was to combine research into advanced materials with digital technology," says Mike Stenson, Kingspan's head of innovation. "In doing that, we realized we could create a building that was both digital and **sustainable** from the ground up."

sustainable /səˈsteməbl/ *adj.*
able to continue for a long time 可持续的

F

"There are multiple sensors throughout IKON that enable us to measure energy consumption," he continues. "We also utilize natural lighting, rainwater, solar panels, and other elements to create a workspace built on sustainable practices."

G

Sustainable **Ambitions**

This focus on sustainability **feeds into** Kingspan's 10-year mission to increase its direct use of **renewable** energy to 60%. In fact, the company aims to achieve net zero carbon manufacturing by 2030.

ambition /æmˈbɪʃn/ *n.*
a strong desire to achieve something 雄心；野心；志气；抱负；志向

renewable /rɪˈnjuːəbəl/ *adj.*
natural resources such as wind, water, and sunlight which are always available （资源）可再生的

H

IKON itself was built using **recycled** materials. A lot of work was also done on the energy modeling of the building during the design stage. "For instance, solar panels on the roof generate enough power to meet 35% of the building's energy demand," Stenson says. "The car park contains charging points for electric cars. There's even a rainwater recovery system for use in toilets."

recycle /rɪːˈsaɪkəl/ *v.*
to put used objects or materials through a special process so that they can be used again 回收利用；循环使用

I

Made from recycled plastic bottles, the building's insulation follows in this path. The company now plans to recycle between 300 and 400 million plastic bottles each year.

J

Digital Inspiration

The center is also home to Kingspan's Digital Team. This team leads research into how the company can bring more intelligence into its products, prepare for Industry 4.0, and **consistent** with the "model first" **approach** to creating new buildings.

consistent /kənˈsɪstənt/ *adj.*
always keeping to the same pattern or style; unchanging 一贯的；前后一致的

approach /əˈprəʊtʃ/ *n.*
way of dealing with a person or thing 方法；手段

K

For the team, IKON provides the perfect setting to ask some of construction's big questions: What role will increase reality, **virtual reality**, and the Internet of Things play in building management? How will **artificial intelligence** and Big Data transform architectural design?

L

Through technology partnerships and its own findings, Kingspan aims to make IKON the world's first level-5 maturity digital twin. Using technologies such as Autodesk Forge, Kingspan uses devices and video cameras to **visualize** building performance in the context of IKON's rich 3D BIM (Building Information Modeling) data, creating an "occupancy-aware" building.

visualize /ˈvɪʒʊəˌlaɪz/ *v.*
to form a picture of someone or something in your mind 想象；形象化

Data-Driven Construction

M

"For instance, we can **anonymously** gather data related to people, behavioral patterns, and time of day," says Brian Glancy, Kingspan's head of BIM strategy. "Through

anonymously /əˈnɒnɪməslɪ/ *adv.*
without giving a name 不具名地；化名地

this, we're better able to **live up to** the idea of performance products for performance environments. **Catering** our buildings **to** the behavior of their occupants is already translating into things like improved insulation performance to gain more space."

N

Glancy says better **command** of building data also means systems such as HVAC and sanitation can be more responsive to human activity. To achieve this, a computer-vision system from Autodesk is used at IKON. Designed from the ground up with anonymity in mind, the system directly addresses data-usage concerns in similar solutions.

command /kəˈmɑːnd/ *n.*
ability to use or control sth.; mastery 使用或控制某事物的能力；掌握

O

By turning human activity into data and analyzing it against metrics such as energy consumption by room or floor, or how people socially occupy a space, Kingspan hopes to help the construction industry create better occupant-simulation models. These could inform the creation of more energy-efficient buildings **optimized** for comfort, wellness, and productivity.

optimize /ˈɒptɪˌmaɪz/ *v.*
to improve the way that something is done or used so that it is as effective as possible 使优化

P

Buildings That Learn

Kingspan's Stenson believes that IKON's novel use of data, **reliance** on green materials,

reliance /rɪˈlaɪəns/ *n.*
confidence or trust in sb./sth.; dependence on sb./sth.（对某人/物的）信任，信赖，信心

and living laboratory structure will continue to drive company and industry innovations for years to come.

Q

The team's research shows that one day, buildings could **take advantage of** machine-learning **algorithms** to respond to behavioral and environmental data in a **proactive** manner. Modern, interconnected building subsystems could maintain comfort as well as productivity for occupants, while minimizing energy use.

proactive /prəʊˈæktɪv/ *adj.*
making things happen or change rather than reacting to events 有前瞻性的，先行一步的；积极主动的

R

IKON's use of data may also advance construction's contribution to the circular economy. With **access** to hyper-detailed building data, companies **dismantling** properties will know exactly which materials and components have been used and where they've been installed. This information makes material recovery easier, turning demolition sites into resources rather than sources of waste.

access /ˈæksɛs/ *n.*
opportunity or right to use sth. or approach sb.（使用某物或接近某人的）机会或权利

dismantle /dɪsˈmæntəl/ *v.*
to take a machine or piece of equipment apart so that it is in separate pieces 拆除

S

"IKON allows us to try out building and design software in a live setting, explore new directions in areas like solar, and improve access control—all done with data," Stenson says. "I think we've only **scratched the surface**."

(Words: 908)

Useful Expressions

account for （数量、比例上）占
feed into 注入，流入；提供原料
live up to 不辜负；做到；实践
cater... to... 满足某种需要或要求
take advantage of 充分利用
scratch the surface 只做了肤浅的研究；不深刻，不周详

Proper Names

Space Needles 太空针塔（美国西雅图著名建筑）
observation tower 瞭望塔
virtual reality 虚拟现实
artificial intelligence 人工智能
algorithm （尤指计算机程序中的）演算法

TEXT A Exercises

1. Content Questions

Each of the following statements contains information given in one of the paragraphs in the TEXT A. Identify the paragraph from which the information is derived. You may choose a paragraph more than once. Each paragraph is marked with a letter.

1) ( ) Construction industry leaders now focus on innovations.

2) ( ) Buildings could take advantage of machine-learning algorithms to respond to behavioral and environmental data.

3) ( ) IKON's use of data may also advance construction's contribution to the circular economy.

4) ( ) Modern, interconnected building subsystems could maintain comfort as well as productivity.

5) ( ) Kingspan aims to make IKON the world's first level-5 maturity digital twin.

6) ( ) It shows that the construction and operation of buildings account for 36% of global energy use.

7) ( ) Better command of building data also means systems such as HVAC and sanitation can be more responsive to human activity.

8) ( ) We're better able to live up to the idea of performance products for performance environments.

2. Vocabulary

A. Fill in the gaps with the words or phrases given in the box. Change the form when necessary.

access	annual	enthusiasm	pavilion	observe
innovation	confirm	account	evolution	represent
commitment	renewable	ambition	consistent	approach

1) He is not fond of ______________.

2) The hotel has exclusive __________to the beach.

3) We made a__________ to keep working together.

4) __________sources of energy must be used.

5) Can you _________what happened?

6) He's covering the party's __________ conference.

7) Even when I was young I never had any_________.

8) I really admire your_________.

9) The carvings __________ a hunting scene.

10) I think it's time we tried a fresh________.

B. Fill in the gaps with the phrases in the box. Change the form when necessary.

account for	feed into	live up to
cater to	take advantage of	scratch the surface

1) They only publish novels which_________ the mass-market.

2) It is surprising that it has taken people so long to________ what is a win-win opportunity.

3) To make an investigation one, should go into matters deeply, not just_________.

4) It is vital that once an institution claims to be particularly good at something, it must ____________ it.

5) We have to__________ every penny we spend on business trips.

3. Translation

Translate the following paragraph into English.

这项研究指导了 Kingspan 全球创新中心 IKON 的设计和建造。IKON 于 2019 年在都柏林北部卡文郡的湖泊和河流中开放，代表着Kingspan对绿色未来的承诺，投资 1000 万欧元。它的功能就像一个“生活实验”，为工程师提供完美的现实世界设置，以衡量新材料的节能性能。

____

____

____

____

____

TEXT B

Digitizing the Eiffel Tower: Inside the World's Largest Urban 3D Model

Since the invention of photography, the Eiffel Tower has been **immortalized** from every angle—a symbol of Paris around the world. Thanks to BIM (Building Information Modeling), this **collective** memory of the Eiffel Tower has now been permanently **captured** in a new medium: the 3D model.

The city of Paris used BIM to immortalize a 133-acre area for "Grand Site **Tour Eiffel**", an international design competition. Created in partnership with Autodesk, the resulting urban model is the largest of its kind in the world. Containing 342 GB of **point-cloud** data, it accurately represents every building, road, tree, and fountain on the site. This model wasn't just

immortalize /ɪˈmɔːtəˌlaɪz/ *v.*
to make someone or something famous for a long time 使永恒；使不朽

collective /kəˈlɛktɪv/ *adj.*
of, by or relating to a group or society as a whole; joint; shared 集体的；共有的

capture /ˈkæptʃə/ *v.*
to succeed in representing (sb./sth.) in a picture, on film, etc 捕捉（画面或影片中的某人/物）

created to immortalize the Eiffel Tower and its surroundings; it was used to **renovate** them.

The competition brief was to modernize the visitor experience of the Eiffel Tower area while respecting its history. Teams of architects, urban planners, engineers, and **landscape** designers from around the globe were all encouraged to submit their designs. Each project needed to be based on a common digital model to ensure all teams could compete on equal **topography**, technical and technological base.

Inside a Virtual Viewpoint

The challenge of creating this urban model for the competition was two-fold. First, **candidates** had to be able to view the landscape from every angle in order to properly **modify** it. Second, projects had to be open to public participation. Each had to offer the jury and Parisians alike a virtual-reality (VR) tour of its **potential** new Eiffel Tower area. This allowed Parisians to provide feedback on the many **proposals** before they went to the jury for a final vote.

Using scan-to-BIM technology, Autodesk worked with French surveying firm Gexpertise for several weeks to map a 3D model of the site. To capture information, Gexpertise used LiDAR, **dispatching** land-based and mobile

renovate /ˈrɛnəˌveɪt/ *v.*
to restore (especially old buildings) to good condition 修复（尤指旧建筑物）；整修

landscape /ˈlændˌskeɪp/ *n.*
everything you can see when you look across an area of land, including hills, rivers, buildings, trees, and plants 风景；景色

topography /təˈpɒgrəfɪ/ *n.*
(description of the) features of a place or district 地形；地志；地形学

candidate /ˈkændɪˌdeɪt/ *n.*
someone who is being considered for a position 候选人

modify /ˈmɒdɪˌfaɪ/ *v.*
change (sth.) slightly, especially to make it less extreme or to improve it 修改，修饰，更改

potential /pəˈtɛnʃəl/ *adj.*
someone or something is capable of developing into the particular kind of person or thing mentioned 潜在的，可能的

proposal /prəˈpəuzəl/ *n.*
a plan or an idea, often a formal or written one, which is suggested for people to think about and decide upon （常为正式书面的）提议

dispatch /dɪˈspætʃ/ *v.*
to send sb./sth. off to a destination or for a special purpose 派遣；发送

laser scanners, and even cameras. The team then used **photogrammetry** to **acquire** a global point cloud representing a complete topography of the area.

"In addition to the site's **heritage** and symbolic nature, the acquisition of data for public use represented an **unprecedented** request—especially considering the scope of the project," says Pauline Barbier, associate director of Gexpertise's modeling division.

A surveyor's precision was particularly valuable. The topography of the 133 acres proved to be extremely complex: Every detail—from gravel paths and fountains to the area's 425 benches, 560 light fixtures, 25 statues, 100 garbage cans, 1,000 buildings, and 8,200 trees and flowerbeds—all had to be recorded. Hundreds of hours of data capture were all processed using Autodesk ReCap. This **resulted in** a massive set of more than 10.3 billion points, or 342 GB of point-cloud data.

The Models Take Shape

Once this work was completed, the **talents** of Gexpertise were combined with those of Canadian engineering firm WSP, a specialist in 3D urban-scale modeling. Processed in Autodesk InfraWorks, the surveys **extracted** nearly 200 point clouds,

acquire /əˈkwaɪə/ *v.*
gain (sth.) by one's own ability, efforts or behaviour 获得，得到

heritage /ˈhɛrɪtɪdʒ/ *n.*
all the qualities, traditions, or features of life there that have continued over many years and have been passed on from one generation to another 遗产；传统

unprecedented /ʌnˈprɛsɪˌdɛntɪd/ *adj.*
never having happened, been done or been known before 无前例的；前所未有的；空前的

talent /ˈtælənt/ *n.*
the natural ability to do something well 天赋；天才

extract /ˈekstrækt/ *v.*
to carefully remove a substance from something which contains it 提取，提炼

which were then used to generate two VR models using Autodesk 3ds Max and Unreal Engine.

"Our initial mission was to produce a model that could be used by the architects and engineers applying for the competition," says Kevin Gilson, manager of design visualization at WSP. "The aim was to help them understand the existing context while also **serving as** a basis for their projects. These had to **take into account** both the landscape and the architectural **constraints** of a listed site.

constraint /kənˈstreɪnt/ *n.*
something that limits or controls what you can do 限制，约束

"I myself visited Paris several times throughout the project," Gilson continues. "I think that even with the accuracy of the data, it is **essential** to breathe the atmosphere of the places you are working on to stay as close as possible to reality." This step provided an opportunity for WSP to collaborate closely with architecture and urban-planning agencies, as well as engineering firms, via Autodesk's BIM 360 platform.

essential /ɪˈsɛnʃəl/ *adj.*
necessary; indispensable; most important 必要的；不可缺少的；最重要的

The Canadian company then **integrated** models provided by the competing teams into the second model, which served as a master model, before generating an interactive 3D version of each of the four projects **in accordance with** the city of Paris's "Discover, Approach, Visit" initiative.

integrate /ˈɪntɪˌgreɪt/ *v.*
to combine sth. in such a way that it becomes fully a part of sth. else （将某事物与另一事物结合）构成整体，融合

The **Countdown** to a New Park Begins

The final step was to present the projects to the community. The public previewed designs and shared feedback on a **dedicated** site. Projects then went to the jury.

Following each competing team's final challenge, which involved oral presentations before the jury, the plan from agency Gustafson Porter + Bowman and its partners proved to be the most **convincing** proposal.

The redevelopment of the area around the Eiffel Tower will take place between 2021 and 2023. The site will remain open to the public throughout this period so the world can **witness** the transformation of one of its most symbolic spaces.

(Words:779)

countdown /ˈkaʊntˌdaʊn/ *n.*
the counting aloud of numbers in reverse order before something happens, especially before a spacecraft is launched 倒计时

dedicated /ˈdedɪkeɪtɪd/ *adj.*
devoted to a cause or ideal or purpose 专注的；献身的

convincing /kənˈvɪnsɪŋ/ *adj.*
making you believe that something is true or right 令人信服的；有说服力的

witness /ˈwɪtnɪs/ *v.*
to be present at (sth.) and see it 当场见到（某事物）；目击

Useful Expressions

result in 产生某种作用或结果

serve as 担任……，充当……；起……的作用

take into account 计及；斟酌；体谅；考虑

in accordance with 按照或依据某事物，一致，符合

Proper Names

Tour Eiffel 埃菲尔铁塔

point-cloud 点云

photogrammetry 照相测量法

TEXT B Exercises

1. Read TEXT B and decide whether the statements are true (T) or false (F).

1) ( ) The Eiffel Tower is a symbol of Paris around the world.

2) ( ) A surveyor's precision was particularly valuable.

3) ( ) The final step was to present the projects to the community.

4) ( ) The challenge of creating this urban model for the competition was three-fold.

5) ( ) Only one project needed to be based on a common digital model.

6) ( ) The redevelopment of the area around the Eiffel Tower will take place between 2021 and 2023.

7) ( ) Using scan-to-BIM technology, Autodesk worked with French surveying firm.

8) ( ) The topography of the 133 acres proved to be extremely complex.

2. Translation

Translate the sentences into English, using the words or phrases in brackets.

1) 这是集体的决定。（collective）

2) 她的油画捕捉住了秋天乡村的微妙色调。（capture）

3) 然而，这并不意味着为了翻新和保护遗产建筑而应打压现代化和新建。（renovate）

4) 这风景仅仅是被一排村庄所破坏。（landscape）

5) 正在物色合适的人选。（candidate）

6) 病人获得有关如何调节自己饮食的指导。（modify）

7) 人们将永远不会明白为什么这些政治问题能获得如此的影响力。（acquire）

8) 这个建筑是我们民族遗产的一部分。（heritage）

9) 他是个了不起的人才。（talent）

10) 金钱对于幸福并非必不可少。（essential）

TEXT C

Forbidden City Architecture

The Forbidden City is the largest **medieval** palace architecture in the world, and was the main **imperial** palace of China's final two **dynasties**: the Ming (1368–1644) and Qing (1616–1911) dynasties.

While most of the buildings in the Forbidden City are made from wood and have a similar style, its architecture is nevertheless rich in subtle **variations** and symbolism. It has some of the grandest and most historically significant buildings and features in China.

It covers a vast area, about 150,000 square meters, and includes about 980 buildings, so it can be hard to know what to **look out for**. To help you make the most of your visit, here are 3 of its architectural **highlights** to **keep an eye out for**.

medieval /ˌmɛdɪˈiːvəl/ *adj.*
of the Middle Ages, about AD 1100-1400 中古的，中世纪的

imperial /ɪmˈpɪərɪəl/ *adj.*
of an empire or its ruler 帝国的；帝王的；皇帝的

dynasty /ˈdɪnəstɪ/ *n.*
a period of time during which a country is ruled by members of the same family 朝代

variation /ˌvɛərɪˈeɪʃən/ *n.*
a change or slight difference in a level, amount, or quantity 变化；差别

look out for 留心，提防；留意找

highlight /ˈhaɪˌlaɪt/ *n.*
best, most interesting or most exciting part of something 最有意思或最精采的部分

keep an eye out for 密切注意；提防

Layout: The South-North **Axis** of Power

The Forbidden City was symmetrically on the north-south central axis of old Beijing.

The south-north axis is one of the most important features of the Forbidden City's layout.

From its main southern entrance through its **majestic** halls to its northern emperors' quarters, the south-north axis was believed to point visitors towards Heaven (the North Star was thought to be Heaven as it is the only seemingly stationary star in the northern sky).The emperor was believed to represent Heaven and was therefore housed in the north.The palace complex is centered on the south-north axis of the old city of Beijing.

Also, Confucian thought is **influential** in the layout; it establishes stability and represents harmony between man and earth. For example, this is why the main gate faces south and the main road to the palace runs on a north-south axis.

According to **Confucianism**, emperors held **supreme** power from Heaven, which allowed them to govern the whole nation. To show this power's central place in the nation: the imperial palace had to be built in the center of the capital city on its south north axis. The important buildings had to be on the

axis /ˈæksɪs/ *n.*
line that divides a regular figure into two symmetrical parts 轴线

majestic /məˈdʒɛstɪk/ *adj.*
having or showing majesty; stately; grand 威严的；壮丽的；高贵的；宏伟的

influential /ˌɪnfluˈɛnʃəl/ *adj.*
having influence; persuasive 有影响的；有说服力的

Confucianism /kənˈfjuːʃənɪzəm/ *n.*
confucianism is a Chinese religious philosophy that emphasizes human morality and correct personal behaviour 儒家；儒学

supreme /sʊˈprɪːm/ *adj.*
highest in authority, rank or degree（权力、级别或地位）最高的；至高无上的

south-north “axis of power”.

Wooden Construction

The Forbidden City’s beams and columns are made of wood, as are the walls that separate the halls into different rooms. Culturally, wood was the favored material in traditional Chinese buildings.

The Forbidden City is the world’s largest collection of well-**preserved** medieval wooden structures. All the buildings in the Forbidden City are made using high quality wooden beams and columns, and there are many examples of outstanding carpentry.

preserve /prɪˈzɜːv/ *v.*
to keep or maintain (sth.) in an unchanged or perfect condition 保护，维护

For instance, its complex interlocking roof brackets, known as dougong, which literally means “cap and block”, not only look impressive; they also have a **crucial** practical application. The brackets **transfer** the weight to the structure’s vertical columns, reducing the pressure on the horizontal beams, which reduces the risk of the beams **splitting**. What is most impressive is that they don’t **require** glue or fasteners; they fit together perfectly because of the quality and precision of the carpentry. It is an innovation that could be up to 2,500 years old.

crucial /ˈkruːʃəl/ *adj.*
very important; decisive 至关重要的；决定性的

transfer / trænsˈfə / *v.*
to move from one place to another 转移

split /splɪt/ *v.*
to break open or apart suddenly 裂开，碎裂

require /rɪˈkwaɪə/ *v.*
to need something 需要，要求

As well as using them for their practicality, architects later focused on making them more decorative, which is very apparent when you look at the complex carpentry of the Forbidden City's roofs.

Painting and **Decorations**

decoration /ˌdɛkəˈreɪʃən/ *n.* something used to beautify 装饰，装潢

Most of the columns in the Forbidden City are painted red. Painting and decoration changed greatly over the Forbidden City's 500-year imperial history, but some things remained **constant**. The windows and doors were often changed by different emperors and their family members to suit their personal needs, and the wall decorations for each hall changed frequently.

constant /ˈkɒnstənt/ *adj.* unchanging; fixed 不变的；恒定的；稳定的

Most of the columns in the Forbidden City are painted red, China's most auspicious color, which gives the entire area a more uniform look. As well as being a decorative feature, the oil paint helps prevent the wood from becoming worse.

Most of the decorations on the buildings can be classified into three types: imperial drawings of dragons and phoenixes, **geometric** patterns, and Suzhou garden patterns.

geometric /ˌdʒɪəˈmɛtrɪk/ *adj.* patterns or shapes consist of regular shapes or lines 几何图形的

Dragons and phoenixes are the major patterns found throughout the Forbidden City. Dragons were used to represent **emperors** while phoenixes represented empresses. The dragons within the Forbidden City, of which there are more than 10,000, are in many

emperor /ˈɛmpərə/ *n.* a man who rules an empire or is the head of state in an empire 皇帝

different styles.

Besides the major buildings, other **pavilions** and towers are decorated with Suzhou garden patterns. The same style of pattern within the Forbidden City can be found in the classical gardens of Suzhou.

(Words: 713)

pavilion /pəˈvɪljən/ *n.*
an ornamental building in a garden or park 亭子

TEXT C Exercise

Translate the following paragraph into Chinese.

Today, concern for the environment balances that enthusiasm for progress. Where they once aimed for Space Needles and flying-saucer-like observation towers, construction industry leaders now focus on innovations that make buildings greener and more responsive to the needs of occupants—thinking smaller, you might say, in order to see the bigger picture.

Part III TASKS

1. Work in pairs and discuss the following questions

1) Do you know any famous architectural designs?

2) Which one do you like best, and why?

3) If your friends from abroad come to visit China, which famous Chinese architecture would you like to recommend? And if you are the guide, what features would you probably introduce to them?

2. Writing

Summarize your answers to the questions above and then write an essay entitled "*My Favorite Architectural Design*". You should write the essay according to the following structures:

1) The basics of my favorite architectural design.

2) Appearance and features of the architectural design.

3) Your reasons for your choice.

When you finish, exchange your writing with a partner and evaluate each other's writing according to the following standards:

1) You must write an essay of at least 120 words.

2) You must use transitional words and phrases to guide readers through your analysis.

3) Try to use as many as possible of new words and expressions in this unit.

Unit 5 Construction Safety

Part I OVERVIEW

In industries like construction, the risk of injury on the job is very real. According to the U.S. Bureau of Labor Statistics, workers in construction and related industries missed work due to injuries and illnesses significantly more often than those in other industries. A construction safety program is crucial for protecting the business's most valuable asset —employees. In many smaller companies, employees are like part of the family, and keeping them happy and safe is vital to the success of the business.

Part II TEXTS

TEXT A

10 Simple Construction Site Safety Rules

A

Construction sites are dangerous places to work. Every year, thousands of people are injured at work on construction sites. So, if you work in construction, it's even more important that you put health and safety into everything you do.

B

Follow these 10 simple construction site safety rules to keep yourself, and others, safe.

C

1. Wear your **PPE** (Personal Protective Equipment) at all times.

When you enter the site, make sure you have the PPE you need. PPE is important, it's your last **line of defense** should you **come into contact with** a hazard on site. Safety boots give you **grip** and protect your feet. Hard hats are easily replaced, but your **skull** isn't. It can't protect you if you don't wear it. Wear your hard hat, safety boots and vest as a minimum, along with any additional PPE required for the task being carried out.

grip /grɪp/ *n.*
an act of holding sb./sth. tightly; a particular way of doing this 紧握；紧抓

skull /skʌl/ *n.*
the bone structure that forms the head and surrounds and protects the brain 颅骨；头（盖）骨

D

2. Do not start work without an **induction**

Each site has its unique hazards and work operations. No two sites are exactly the same. Make sure you know what is happening so that you can work safely. Inductions are a legal requirement on every construction site you work on. Your induction is important. It tells you where to sign in, where to go, what to do, and what to avoid. Don't start work without one.

induction /ɪn'dʌkʃ(ə)n/*n.*
the process of introducing sb. to a new job, skill, organization, etc.; a ceremony at which this takes place 就职；入门；接纳会员；就职仪式

E

3. Keep a tidy site

Construction work is messy. **Slips** and trips might not seem like a major problem **compared to** other high-risk work happening

slip /slɪp/ *n.*
act of accidentally sliding and losing your balance 滑；滑倒

on the site, but don't be fooled. According to **HSE** statistics, slips and trips accounted for 30% of specified major injuries on construction sites. Remember to keep your work area tidy throughout your shift to reduce the number of slip and trip hazards. **Pay** particular **attention to** areas such as access and escape routes.

F

4. Do not put yourself or others **at risk**

Actions speak louder than words. Especially on construction sites where one wrong move could put you in harm's way. Set a good example, think safe and act safely on site. You are responsible for your own behavior. Construction sites are dangerous places to work. Make sure you remain safety aware throughout your shift.

G

5. Follow safety signs and procedures

Follow construction safety signs and procedures. These should be explained to you in your induction (rule number 2). Your employer should ensure a risk **assessment** is **carried out** for your activities. Make sure you read and understand it. Control measures are put in place for your safety. Make sure they are **in place** and working before you start.

assessment /ə'sesmənt/ *n.*
an opinion or a judgement about sb./sth. that has been thought about very carefully 看法；评估

H

6. Never work in unsafe areas

Make sure your work area is safe. Know what is happening around you. According to HSE statistics, 14% of fatalities in construction were caused by something **collapsing** or **overturning,** and 11% by being struck by a moving vehicle. Don't work at height without suitable **guard rails** or other fall prevention. Don't enter unsupported trenches. Make sure you have safe access. Don't work below **crane loads** or other dangerous operations.

I

7. Report defects and near misses

If you notice a problem, don't **ignore** it, report it to your supervisor immediately. Fill out a **near-miss report**, an incident report, or simply tell your supervisor. Whatever the procedure in place on your site for reporting issues, use it. Action can only be taken quickly if the management has been made aware of the problem. The sooner problems are resolved the less chance for an accident to occur.

J

8. Never tamper with equipment

If somethings not working, or doesn't look right, follow rule number 7 and report it. Don't try and force something, or **alter**

collapse /kə'læps/ *v.*
to fall down or fall in suddenly, often after breaking apart （突然）倒塌，坍塌

overturn /əʊvə'tɜːn/ *v.*
turns upside down or on its side 打翻；倾覆；翻掉

ignore /ɪg'nɔː/ vt.
to pay no attention to sth. 忽视；对……不予理会

alter /'ɔːltə/ *v.*
to become different; to make sb./sth. different （使）改变，更改，改动

something, if you're trained to or supposed to.

Never remove guard rails or scaffold ties. Do not remove machine guards. Do not attempt to fix **defective** equipment unless you are **competent** to do so. Do not ever tamper with equipment without **authorization.**

defective /dɪ'fektɪv/ *adj.*
having a fault or faults; not perfect or complete 有缺点的；有缺陷的；有毛病的

competent /'kɒmpɪt(ə)nt/ *adj.*
having enough skill or knowledge to do sth. well or to the necessary standard 足以胜任的；有能力的；称职的

authorization /ɔːθəraɪ'zeɪʃ(ə)n/ *n.*
official permission or power to do sth.; the act of giving permission 批准；授权

K

9. Use the right equipment

One tool does not fit all. Using the correct tool for the job will get it done quicker, and most importantly, safer. Visually check equipment is in good condition and safe to use before you start. Only use 110v equipment on the site. 240v equipment is strictly prohibited without prior authorization from management and will only be used if no 110v **alternative** available and additional safety **precautions** are taken.

alternative /ɔːl'təːnətɪv/ *n.*
a thing that you can choose to do or have out of two or more possibilities 可供选择的事物

precaution /prɪ'kɔːʃən/ *n.*
something that is done in advance in order to prevent problems or to avoid danger 预防措施；预防；防备

L

10. If in doubt, ask

Unsure what to do? Or how to do something safely? Or you think something is wrong? Stop work, and ask. It takes 5 minutes to check, but it might not be so easy to put things right if things go wrong. It's better to be safe than sorry. Mistakes on construction sites can cost lives, don't let it be yours.

(words: 799)

Useful Expressions

come into contact with 接触到；联系；开始做某事

compare to 把……比作，比喻为

pay attention to 注意；重视

at risk 处于危险中

carry out 执行，实行；贯彻；实现；完成

in place 适当，适当的；在适当的地方，在恰当的位置

Proper Names

PPE (Personal Protective Equipment) 个人防护用品

HSE (Health and Safety Executive) （英国）卫生安全管理局

line of defense 防线、防卫线

guard rail 防护轨

crane load 吊车荷载；起重机起重量，吊车负荷

near-miss report 未遂事故报告

TEXT A Exercises

1. Content Questions

Each of the following statements contains information given in one of the paragraphs in the TEXT A. Identify the paragraph from which the information is derived. You may choose a paragraph more than once. Each paragraph is marked with a letter.

1) ( ) PPE is important, it's your last line of defense should you come into contact with a hazard on site.

2) ( ) Every year, thousands of people are injured at work on construction sites.

3) ( ) Focus on areas such as access and escape routes.

4) ( ) The sooner problems are resolved the less chance for an accident to occur.

5) ( ) Do not attempt to fix defective equipment unless you are competent to do so.

6) ( ) Visually check equipment is in good condition and safe to use before you start.

7) ( ) Your employer should ensure a risk assessment is carried out for your activities.

8) ( ) Make sure you know what is happening so that you can work safely.

2. Vocabulary

A. Fill in the gaps with the words or phrases given in the box. Change the form when necessary.

grip	skull	induction	assessment	overturning
collapse	ignore	defective	competent	alter
alternative	slip	precaution	authorization	

1) You can be paid in cash weekly or by cheque monthly; those are the two________.
2) Make sure the firm is___________ to carry out the work.
3) Prices did not___________ significantly during 2004.
4) He ___________all the "No Smoking" signs and lit up a cigarette.
5) Objective __________of the severity of the problem was difficult.
6) The roof_________ under the weight of snow.
7) She_________ over on the ice and broke her leg.
8) Can I see your__________?
9) Her hearing was found to be slightly________.
10) So we are going to deal with _________this week.

B. Fill in the gaps with the phrases in the box. Change the form when necessary.

carry out	pay attention to	at risk
in place	compare to	come into contact with

1) She needed a clear head to________ her instructions.
2) He was putting himself__________.
3) We already have our core team_________.
4) I didn't _________what she was saying.
5) A key process in interpersonal interaction is that of social comparison, in that we evaluate ourselves in terms of how we___________others.

3. Translation

Translate the following paragraph into English.

各地区、有关部门和国有企业必须清楚地认识到建筑业面临的严峻安全形势。要增强安全红线意识，全面提高建筑业安全水平，切实履行承建商责任，加强风险管理和执法，特别是在地质复杂、洪水风险高等的危险区域。

____

____

____

____

____

____

TEXT B

The History of Safety in a Construction Environment

In 2016, 10.3 million U.S. workers were employed in the construction industry. That means that construction jobs account for about 5% of all of the United States **workforce**, but construction workers account for more than 17% of workplace **fatalities** in the American workforce.

workforce /ˈwɜːkfɔːs/ *n.*
all the people who work for a particular company, organization, etc. 劳动力；工人总数，职工总数

fatality /fəˈtælətɪ/ *n.* (pl. -ies)
a death that is caused in an accident or a war, or by violence or disease （事故、战争、疾病等中的）死亡

With so many laws and safety regulations in place, it's hard to believe how dangerous these jobs still are. By looking at the history of safety in a construction environment, we can see that there's been a **definite** improvement in workplace safety; and we can also see that it hasn't been an easy journey to develop the standards that we have in place today.

definite /ˈdefɪnət/ *adj.*
~ (that...) sure or certain; unlikely to change 肯定的；确定的；不会改变

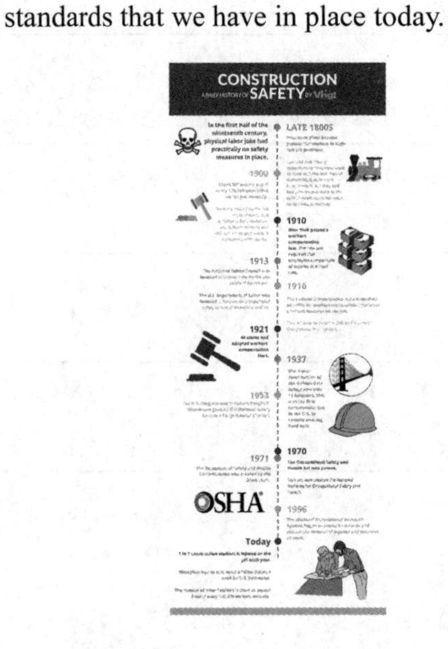

Late 1800s

In the first half of the nineteenth century, physical labor jobs had practically no safety measures in place. We're starting later in the century because this is when things began changing for the better. **Post Civil War** life, workers began **setting aside** money and purchasing **insurance** in case of an accident on the job. Some employers even began providing insurance plans for their employees

insurance /ɪnˈʃʊərəns/ *n.*
~ (against sth.) an arrangement with a company in which you pay them regular amounts of money and they agree to pay the costs, for example, if you die or are ill/sick, or if you lose or damage sth. 保险

or providing alternative jobs for their injured men. Some workers left jobs that they felt were too dangerous, and many employers had to raise wages on high-risk jobs to **attract** enough workers. It was this climate that began to **influence** changes in industry policies.

Railroad and mining regulatory **commissions** were formed with the **intention** of mandating a safer work environment, but they had few powers and were rarely able to exert much influence on working conditions.

1900s

In the year 1900, about three hundred miners out of every one hundred thousand were killed on the job annually. To compare, that number is around nine out of every one hundred thousand **annually** today.

At this point in time, workers injured on the job would have to sue employers for damages—and actually winning was difficult. Only about half of all workplace fatalities resulted in **compensation** for the family, and even then the amount was only **equal to** about half a year's pay. With such a low cost to employers for employee deaths and **virtually** no legal **consequences,** safety was of little concern in the workplace.

1910s

Following European examples, New York

attract /əˈtrækt/ *v.*
/ usually passive / ~ sb. (to sb./sth.) if you are attracted by sth., it interests you and makes you want it; if you are attracted by sb., you like or admire them 吸引；使喜爱；引起……的好感（或爱慕）

influence /ˈɪnfluəns/ *n.*
to have an effect on the way that sb. behaves or thinks, especially by giving them an example to follow 影响；起作用

commission /kəˈmɪʃn/ *n.*
(often Commission) [C] an official group of people who have been given responsibility to control sth., or to find out about sth., usually for the government（通常为政府管控或调查某事的）委员会

intention /ɪnˈtenʃn/ *n.*
~ (of doing sth.)~ (to do sth.)~ (that...) what you intend or plan to do; your aim 打算；计划；意图；目的

annually /ˈænjuəlɪ/ *adv.*
once a year 一年一次地

compensation /ˌkɒmpenˈseɪʃn/ *n.*
something, especially money, that sb. gives you because they have hurt you, or damaged sth. that you own; the act of giving this to sb. 补偿（或赔偿）物；（尤指）赔偿金，补偿金；赔偿

virtually /ˈvɜːtʃuəlɪ/ *adv.*
almost or very nearly, so that any slight difference is not important 几乎；差不多；事实上；实际上

consequence /ˈkɒnsɪkwəns/ *n.*
~ (for sb./sth.) a result of sth. that has happened 结果；后果

passed a workers' compensation law in 1910. Instead of requiring injured workers to **sue for** damages in court, the new law required that employers compensate all injuries at a fixed rate. For workers, this meant better and more **reliable** benefits. For employers, this meant more satisfied employees and more **predictable** costs. By 1921, all except six states had adopted workers' compensation laws.

reliable /rɪˈlaɪəbl/ *adj.*
that can be trusted to do sth. well; that you can rely on 可信赖的；可依靠的

predictable /prɪˈdɪktəbl/ *adj.*
if sth. is predictable, you know in advance that it will happen or what it will be like 可预见的；可预料的

The National Safety Council was founded in 1913 to promote the health and safety of Americans. Recognizing the importance of this effort, the U.S. Congress and **President Dwight D. Eisenhower** granted the **NSC** a Congressional charter.

The U.S. Department of Labor, which was also founded in 1913, focuses on occupational safety as one of its main branches.

In 1916, the ***Federal Compensation Act*** established benefits for workers who sustain injuries or contract illnesses on the job. This act also created the **Office of Workers' Compensation Programs.**

1930s

In more than four years of construction, there were only eleven workplace fatalities during the construction of the **Golden Gate**

Bridge. Ten fatalities were caused by a single incident when a **suspended** platform broke, meaning there were only two fatal incidents during the entire four years of construction. The chief engineer on the project was so concerned with workers' safety that he spent $130,000 on safety nets and he was responsible for the first construction site in the United States to require wearing hard hats. The safety nets alone saved nineteen lives.

suspend /sə'spɛnd/ *v.*
If you suspend something, you delay it or stop it from happening for a while or until a decision is made about it 暂停

1970s

The Occupational Safety and Health Act was passed in 1970.

An Act to assure safe and healthful working conditions for working men and women; by authorizing **enforcement** of the standards developed under the Act; by assisting and encouraging the States in their efforts to assure safe and healthful working conditions; by providing for research, information, education, and training in the field of occupational safety and health; and for other purposes.

enforcement /ɪn'fɔːsmənt/ *n.*
If someone carries out the enforcement of an act or rule, they enforce it 执行

This act also created the **National Institute for Occupational Safety and Health**, which conducts research and makes safety **recommendations.**

recommendation /ˌrekəmen'deɪʃn/ *n.*
~ (to sb.) (for/on/about sth.) an official suggestion about the best thing to do 正式建议；提议

The Occupational Safety and Health Administration was created by the OSHA

(Act) in 1971, to "assure safe and healthful working conditions for working men and women by setting and enforcing standards and by providing training, outreach, education and assistance". OSHA's workplace safety inspections have been shown to reduce injury rates and injury costs without negative effects to employment, sales, credit ratings, or company survival.

1990s – Today

The National Occupational Research Agenda began in 1996 and works to conduct research and reduce the number of injuries and illnesses at work. It was created by the NIOSH to provide a research framework for many different organizations in **collaboration.**

collaboration /kəˌlæbəˈreɪʃən/ *n.*
the act of working together to produce a piece of work, especially a book or some research（尤指著书或进行研究时的）合作

Today, one in seven construction workers is injured on the job each year. Workplace injuries cost about a billion dollars a week for U.S. businesses.

The importance of construction workers in our society is great. They're responsible for our roads, houses, businesses, and maintenance of our country's physical infrastructure. Construction spending in 2016 was **estimated** at around 1.2 trillion dollars. With this job comes many hazards, as we can see from past and present research. We've come a long way with safety in a construction

estimate /ˈestɪmət/ *v.*
a judgement that you make without having the exact details or figures about the size, amount, cost, etc. of sth. （对数量、成本等的）估计；估价

environment in the past hundred years or so, but we are still continuing to improve in an effort to **eliminate** workplace fatalities. Change doesn't come easy, but we can keep working to bring about change to keep workers safe on the job.

(Words:927)

eliminate /ɪˈlɪmɪneɪt/ *v.*
~ sth./sb. (from sth.) to remove or get rid of sth./sb. 排除；清除；消除

Useful Expressions

set aside 省出；抽出
be equal to 等于；相当于；等同；相当
sue for 控告

Proper Names

Post Civil War 内战后
The National Safety Council (NSC) 国家安全委员会
President Dwight D. Eisenhower 德怀特 • D.艾森豪威尔总统
The U.S. Department of Labor 美国劳工部
Federal Compensation Act 联邦补偿法
Office of Workers' Compensation Program 工人补偿计划办公室
Golden Gate Bridge 金门大桥
The Occupational Safety and Health Act 职业安全与健康法
National Institute for Occupational Safety and Health 国家职业安全和健康研究所
The National Occupational Research Agenda 国家职业研究议程

TEXT B Exercises

1. Read TEXT B and decide whether the statements are true (T) or false (F).

1) ( ) In 2019, 10.3 million U.S. workers were employed in the construction industry.

2) ( ) Construction workers account for more than 17% of workplace fatalities in the American workforce.
3) ( ) In the year 1900, about two hundred miners out of every one hundred thousand were killed on the job annually.
4) ( ) There were only eleven workplace fatalities during the construction of the Golden Gate Bridge.
5) ( ) Construction spending in 2018 was estimated at around 1.2 trillion dollars.
6) ( ) The U.S. Department of Labor was founded in 1913,
7) ( ) In the year 1900, only about half of all workplace fatalities resulted in compensation for the family.
8) ( ) The National Safety Council was founded in 1913 to promote the health and safety of Americans.

2. Translation

Translate the sentences into English, using the words or phrases in brackets.

1) 有几个人受伤，但没有人死亡。(fatality)
2) 你最晚明天能给我一个确定的答复吗？(definite)
3) 她首先吸引我的是她的幽默感。(attract)
4) 电视对儿童究竟有什么影响？(influence)
5) 今年的结果几乎和去年的一样。(virtually)
6) 这项决定可能对该行业造成严重后果。(consequence)
7) 我们在物色可靠而又勤奋的人。(reliable)
8) 我可以粗略估计一下你所需要的木材量。(estimate)
9) 有了信用卡就用不着携带很多现金。(eliminate)
10) 那本书的结局完全是可以预见的。(predictable)

TEXT C

Chinese Construction Firms Using AI to Monitor Workers' Safety

An artificial intelligence system tracks workers on a construction site

Artificial intelligence is being used to **monitor** workers' behavior on construction sites across China, according to the Chinese Academy of Sciences.

The technology, developed by the **Institute of Automation** in Beijing, has "made construction sites transparent" with significant improvements in safety and productivity, according to a report published on the academy's website last month.

The report said the AI was hooked up to CCTV cameras and was able to tell whether an employee was doing their job or "**loitering**". It can also **distinguish** between different types

artificial /ˌɑːtɪˈfɪʃl/ *adj.*
人造的；仿造的；虚伪的；非原产地的；武断的

intelligence /ɪnˈtelɪdʒəns/ *adj.*
the ability to learn, understand and think in a logical way about things; the ability to do this well 智力；才智；智慧

monitor /ˈmɒnɪtə(r)/ *n.*
a television screen used to show particular kinds of information; a screen that shows information from a computer 显示屏；监视器；（计算机）显示器

Institute of Automation 自动化研究所

loiter /ˈlɔɪtə/ *v.*
walk up and down without any real purpose 闲逛

distinguish /dɪˈstɪŋgwɪʃ/ *v.*
~ (between) A and B; ~ A from B to recognize the difference between two people or things 区分；辨别；分清

of activity such as smoking or using a smartphone.

The technology also sends alerts when accidents or safety risks, such as a worker forgetting to wear a helmet or entering a restricted area, are identified, and also tracks people involved in forbidden activities such as fighting.

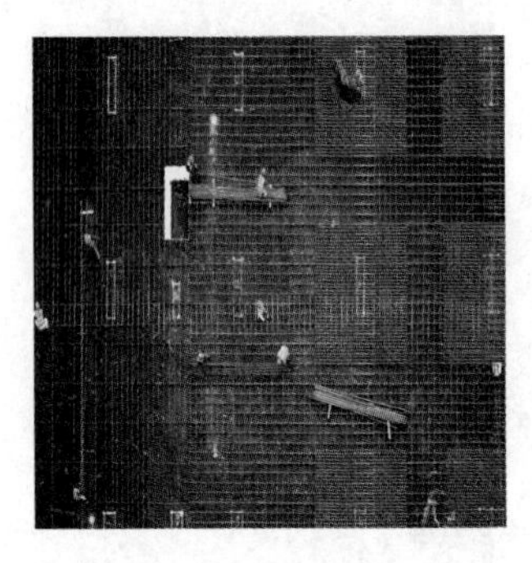

Construction sites are vast **jigsaws** of people and parts that must be pieced together just so at just the right times. As projects get larger, mistakes and delays get more expensive.

jigsaw /ˈdʒɪgsɔː/*n.*
(also "jigsaw puzzle") a picture printed on cardboard or wood, that has been cut up into a lot of small pieces of different shapes that you have to fit together again 拼图；拼板玩具

AI is starting to change various aspects of construction, from design to self-driving diggers. Some companies even provide a kind of overall AI site inspector that matches images taken on site against a **digital** plan of the building. Now British-Israeli startup Buildots is making that process easier than ever by using video footage from GoPro cameras mounted on the hard hats of workers.

digital /ˈdɪdʒɪtl/ *adj.*
using a system of receiving and sending information as a series of the numbers one and zero, showing that an electronic signal is there or is not there 数字信息系统的；数码的；数字式的

When managers tour a site once or twice

a week, the camera on their head captures video **footage** of the whole project and uploads it to image recognition software, which compares the status of many thousands of objects on site—such as electrical sockets and bathroom fittings—with a digital replica of the building.

footage /ˈfʊtɪdʒ/ *n.*
[U] part of a film showing a particular event （影片中的）连续镜头，片段

The AI also uses the video feed to track where the camera is in the building to within a few centimeters so that it can identify the exact location of the objects in each frame. The system can track the status of around 150,000 objects several times a week. For each object the AI can tell which of three or four states it is in, from not yet begun to fully installed.

Site **inspections** are slow and **tedious**, says Sophie Morris at Buildots, a civil engineer who used to work in construction before joining the company. The Buildots AI gets rid of many repetitive tasks and lets people focus on important decisions. "That's the job people want to be doing—not having to go and check if the walls have been painted or if someone's drilled too many holes in the ceiling," she says.

inspection /ɪnˈspekʃn/ *n.*
an official visit to a school, factory, etc. in order to check that rules are being obeyed and that standards are acceptable 视察

tedious /ˈtiːdiəs/ *adj.*
lasting or taking too long and not interesting 冗长的；啰唆的；单调乏味的；令人厌烦的

Another plus is the way the tech works in the background. "It captures data without the need to walk the site with **spreadsheets** or

spreadsheet /ˈspredʃiːt/ *n.*
a computer program that is used, for example, when doing financial or project planning. You enter data in rows and columns and the program calculates costs, etc. from it （计算机）电子表格程序

schedules," says Glen Roberts, operations director at Wates. He says his firm is now planning to roll out the Buildots system at other sites.

Comparing the complete status of a project with its digital plan several times a week has also made a big difference during the covid-19 **pandemic.** When construction sites were shut down to all but the most essential on-site workers, managers on several Buildots projects were able to keep tabs on progress remotely.

pandemic /pænˈdemɪk/ *n.*
a disease that spreads over a whole country or the whole world （全国或全球性）流行病；大流行病

But AI won't be replacing those essential workers anytime soon. Buildings are still built by people. "At the end of the day, this is a very labor-driven industry, and that won't change," says Morris.

(Words: 558)

TEXT C Exercise

Translate the following paragraph into Chinese.

The AI also uses the video feed to track where the camera is in the building to within a few centimeters so that it can identify the exact location of the objects in each frame. The system can track the status of around 150,000 objects several times a week. For each object the AI can tell which of three or four states it is in, from not yet begun to fully installed.

____

____

____

____

____

Part III TASKS

1. Work in pairs and discuss the following questions based on the table.

A survey lists four aspects of effects of accidents on construction sites including "order of production and operations", "reputation of firms", "psychology of labors", and others. Table 1 lists the respondents' views.

Table 1 Effects of accidents on construction sites

Effects of site accidents	Number of firms	Proportion(%)
Order of production and operations	6	30
Preputation of firms	14	70
Psychology of labors	0	0
Others	0	0
Total	20	100

1) Can you illustrate the table in detail?
2) What information does it reflect?
3) Why does the majority of respondents choose "reputation of firms" ? Or why do they think that the significant effect of sit accidents is on reputation of construction firms?
4) Do you have any ideas about it?

2. Writing

Summarize your answers to the questions above. Write a report for a construction firm describing the information shown in the table. You should write at least 150 words.

Unit 6　Civil Engineer

Part I　OVERVIEW

Civil engineers design and construct major transportation projects and play a pivotal role in the effective execution of all manner of engineering projects. They conceive, design, build, supervise, operate, construct and maintain infrastructure projects and systems in the public and private sector, including roads, buildings, airports, tunnels, dams, bridges, and systems for water supply and sewage treatment. Their input, and leadership where necessary is essential to secure the smooth execution of a vast selection of projects.

Part II　TEXTS

TEXT A

Civil Engineers **Overview**

A

What is a Civil Engineer?

B

Civil engineers are responsible for the world's most **inspiring** buildings, bridges and roads, as well as some of the less exciting – yet no less structurally sound – buildings, bridges and roads. Some of these **incredible**

overview /ˈəʊvəˌvjuː/ *n.*
a general understanding or description of a subject or situation as a whole 概述

inspiring /ɪnˈspaɪərɪŋ/ *adj.*
giving people a feeling of excitement and a desire to do sth. great 鼓舞人心的；启发灵感的

incredible /ɪnˈkrɛdəbəl/ *adj.*
difficult to believe; amazing or fantastic 难以置信的；不可思议的

structures include the **Burj Khalifa** in Dubai, the Golden Gate Bridge in San Francisco and "**Spaghetti Junction**" in Birmingham, England, which wouldn't be possible without talented civil engineers.

C

According to the American Society of Civil Engineers, the scope of the **profession** is "the design and maintenance of public works such as roads, bridges, water and energy systems as well as public facilities like ports, railways and airports". Civil engineering **dates back** centuries and is one of the largest sectors in the engineering field. Civil engineers are involved with these projects at every step, from the beginning designs to the construction to the **oversight** when the project is complete. Job responsibilities include analysis – especially in the planning stage – of survey reports and maps. A civil engineer's duties may also include **breaking down** construction costs and considering government regulations and potential environmental **hazards**. Civil engineers also perform experiments, whether that's testing soils to determine the strength of a project's foundation or **assessing** building materials to determine whether, **for instance**, concrete would work best for a certain project. They also need financial ability, since they

profession /prəˈfɛʃən/ *n.*
a type of job that requires advanced education or training 职业

oversight /ˈəʊvəˌsaɪt/ *n.*
management by overseeing the performance or operation of a person or group 监督，照管

hazard /ˈhæzəd/ *n.*
(thing that can cause) danger; risk 危险，风险

assess /əˈsɛs/ *v.*
to place a value on; judge the worth of something 评估，估算

provide cost estimates for equipment and labor, as well as knowledge of software programs for use in planning and designing systems and structures.

D

Civil engineering is a broad field. Specialties include architectural, structural, transportation, traffic, water resources and **geotechnical** engineering. Environmental engineering is another specialty, and it uses engineering principles to both protect the health of nature and people. Civil engineers may work for state or local governments or in the private sector at **consulting** or construction firms. Some civil engineers hold supervisory or administrative positions, while others **pursue** careers in design, construction or teaching. The Bureau of Labor Statistics projects 6.3 percent employment growth for civil engineers between 2018 and 2028. In that period, an estimated 20,500 jobs should open up.

E

What is the Job Like?

F

For the most part, jobs in civil engineering are project-based. While some civil engineers will spend their working hours at the office, others will spend a lot of time in the field. However, many jobs offer a **blend** of

geotechnical /ˌdʒiːəʊˈtɛknɪkəl/ *adj.*
relating to the application of technology to engineering problems caused by geological factors 岩土工程技术的

consult /kənˈsʌlt/ *v.*
to get or ask advice from 商量；向……请教

pursue /pəˈsjuː/ *v.*
to continue doing an activity or trying to achieve something 追求；努力实现

blend /blɛnd/ *v.*
to mix together different elements 混合；协调

field and office work. Government jobs have **predictable** hours, while construction work in the private sector tends to run from 7 a.m. to 3 p.m. Working for a commercial real estate firm could take longer hours and travel. If you're in a managerial position, directing projects might require longer hours as well.

G

How to Become a Civil Engineer?

H

To get an entry-level position, you'll need at least a **bachelor**'s degree in civil engineering. However, according to Shelley Okimoto, graduate advisor for the University of California – Berkeley's School of Civil and Environmental Engineering, an engineer with just a Bachelor of Science may do more of the basic work and could **end up** hitting a salary **ceiling**. She writes in an email: "[Master of Science] graduates often have more responsibility, leading teams and designing projects or creating policy and handling finances. Employers like to have both types around, but finding an employee with stronger training in **critical** thinking skills always helps, and an advanced degree certainly **indicates** more training in that area." According to the BLS, one of every five civil engineers has a master's degree. While in

predictable /prɪˈdɪktəbəl/ *adj.*
capable of being foretold 可预见的

bachelor /ˈbætʃələ/ *n.*
person who holds a first university degree 获学士学位的人

ceiling /ˈsiːlɪŋ/ *n.*
an upper limit on what is allowed 上限，天花板

critical /ˈkrɪtɪkəl/ *adj.*
marked by a tendency to find and call attention to errors and flaws 批评的，批判的

indicate /ˈɪndɪˌkeɪt/ *v.*
to be a sign of (sth.); suggest the possibility or probability of 象征；表明；暗示

graduate school, Okimoto **recommends** developing good study skills, joining a study group and taking advantage of faculty and teaching assistant office hours.

I

Civil engineers also need to **obtain** an engineering license, which involves completing the necessary coursework, getting several years of practical experience and passing the national Fundamentals of Engineering exam, among other requirements.

（words: 636）

recommend /ˌrɛkəˈmɛnd/ *v.*
to suggest (a course of action, treatment, etc); advise 建议；劝告

obtain /əbˈteɪn/ *v.*
to get sth.; come to own or possess sth. 获得

Useful Expressions

date back 追溯到；回溯至

break down 分解

for instance 例如

end up 到达或来到某处；达到某状态或采取某行动

Proper Names

Burj Khalifa 哈利法塔

spaghetti junction 复式公路枢纽；多层式立交桥

TEXT A Exercises

1. Content Questions

Each of the following statements contains information given in one of the paragraphs in the TEXT A. Identify the paragraph from which the information is derived. You may choose a paragraph more than once. Each paragraph is marked with a letter.

1) ( ) Civil engineers are responsible for the world's most inspiring buildings.

2) ( ) You'll need at least a bachelor's degree in civil engineering.
3) ( ) Government jobs have predictable hours, while construction work in the private sector tends to run from 7 a.m. to 3 p.m.
4) ( ) Civil engineering is one of the largest sectors in the engineering field.
5) ( ) According to the BLS, one of every five civil engineers has a master's degree.
6) ( ) Civil engineers also need to obtain an engineering license.
7) ( ) Civil Engineering has many branches.
8) ( ) Civil Engineering include architectural, structural, transportation, traffic, water resources and geotechnical engineering.

2. Vocabulary

A. Fill in the gaps with the words or phrases given in the box. Change the form when necessary.

inspiring	facility	oversight	hazard	assess
incredible	consult	maintenance	instance	geotechnical
pursue	bachelor	indicate	obtain	critical

1) Growing levels of pollution represent a serious health______ to the local population.
2) Have you __________your lawyer about this?
3) She wishes_________ to a medical career.
4) The supervisor is always very_________.
5) Record profits in the retail market _________a boom in the economy.
6) I finally managed to__________ a copy of the report.
7) It seemed_________ that she had been there a week already.
8) The hotel has special_________ for welcoming disabled people.
9) The book is less than___________.
10) It's difficult to___________ the effects of these changes.

B. Fill in the gaps with the phrases in the box. Change the form when necessary.

take for granted	as well as	concerned with
date back	for instance	end up

1) The issue is not a new one. It ___________to the 1930s at least.

2) I think I've been very selfish. I've been mainly _______ myself.

3) Coursework is taken into account _________ exam results.

4) If you smell something unusual (gas fumes or burning,___________), take the car to your mechanic.

5) You __________with a mishmash of policies rather than a consistent national approach.

3. Translation

Translate the following paragraph into English.

在大多数情况下，土木工程的工作都是基于项目的。虽然一些土木工程师将他们的工作时间花在办公室里，但其他人将花很多时间在外地。也有许多工作提供了实地工作和办公室工作。政府部门的工作时间是可预测的，而私营单位的工作往往是从早上 7 点到下午 3 点。在商业房地产公司工作可能需要更长的工作时间和更多出差要求。如果你处于管理职位，也可能需要更多时间指导项目。

____

____

____

____

____

____

TEXT B

Collaborative Action Called for on World Engineering Day

The world will need more equitable, **collaborative** and **interdisciplinary** engineering solutions to address challenges in achieving sustainable development goals, according to a reported published on Thursday by UNESCO and its partners.

collaborative /kəˈlæbərətɪv/ *adj.*
a job or piece of work that involves two or more people working together to achieve something 合作的，协作的

interdisciplinary /ˌɪntəˈdɪsɪˌplɪnərɪ/ *adj.*
involving more than one academic subject 跨学科的

The report was **issued** in celebration of the 2021 World Engineering Day for Sustainable Development on Thursday. It is the second report of its kind in a decade, providing **comprehensive** and authoritative **insights** on the engineering innovations that are shaping the world, as well as challenges in building engineering capacities, especially in developing countries.

issue /ˈɪʃjuː/ *v.*
to send sth. out; make sth. known 发出；颁布；公布

comprehensive /ˌkɒmprɪˈhɛnsɪv/ *adj.*
including all the necessary facts, details, or problems 全面的；综合的

insight /ˈɪnˌsaɪt/ *n.*
ability to see into the true nature (of sth.); deep understanding 洞察力

World Engineering Day was created by UNESCO in 2019 to **celebrate** engineering achievements in advancing sustainable development, but also to raise social awareness for engineers and their profession, so that the younger generation, especially young women, could be interested in pursuing engineering as a career and become leaders in their respective fields.

celebrate /ˈsɛlɪˌbreɪt/ *v.*
to mark (a happy or important day, etc) with festivities and rejoicing 庆祝；祝贺

The report was prepared in collaboration with the **Chinese Academy of Engineering**, the International Centre for Engineering Education based at Tsinghua University, the World Federation of Engineering Organizations and other international engineering organizations.

Li Xiaohong, the president of the Chinese Academy of Engineering, said that the global engineering **community** should do more to achieve the United Nations 2030 Agenda for Sustainable Development, as well as respond more effectively to global challenges including climate change, natural disasters, public health and food security.

community /kəˈmjuːnɪtɪ/ *n.*
a group of people who are similar in some way 团体

"The engineering community has an **undeniable** responsibility," he said. "Whether these goals can be achieved is directly related to the development of engineering technology, which also requires **joint** efforts from the international engineering community."

undeniable /ˌʌndɪˈnaɪəbəl/ *adj.*
that cannot be disputed or denied; undoubtedly true 不可否认的；确定无疑的

joint /dʒɔɪnt/ *adj.*
shared by or belonging to two or more people 联合的；共同的

However, Li said there is still an unbalanced **distribution** of engineering science and technology resources around the world. Innovative and high-quality engineering education also needs to be more **accessible**.

distribution /ˌdɪstrɪˈbjuːʃən/ *n.*
the act of sharing things among a large group of people in a planned way 分布；分配

accessible /əkˈsɛsəbəl/ *adj.*
easy to obtain or use 易使用的；易得到的

"Engineering education equality and capacity building support in underdeveloped countries is getting more **urgent** by the day," he said. "A global engineering community featuring equality, inclusiveness, **diversity** and win-win cooperation needs to be formed."

Gong Ke, WFEO president, said interdisciplinary, inter-sector and international collaboration are needed to invest in and **strengthen** the capacity of engineering innovations across the world, especially in developing countries, to **tackle** some of the world's most urgent problems.

These efforts can help "shape a peaceful, **prosperous**, **inclusive** and sustainable world for all people with no one left behind", he said.

The report provides examples of engineering innovations and practices that are helping to solving key global challenges and are contributing to achieving the sustainable development goals. However, it also highlights some issues in improving engineering capability, such as **transforming** engineering education to suit current needs and a lack of gender diversity in the engineering field.

When discussing details of the report, Marlene Kanga, former president of the **WFEO,** said engineers are essential for sustainable water supply and **sanitation** systems for the world, one of the UN sustainable development goals.

urgent /ˈɜːdʒənt/ *adj.*
needing immediate attention, action or decision 紧急的；迫切的

diversity /daɪˈvɜːsɪtɪ/ *n.*
the fact of including many different types of people or things 多样性

strengthen /ˈstrɛŋθən/ *v.*
to become stronger or make something stronger 加强，巩固

tackle /ˈtækəl/ *v.*
to try to deal with a difficult problem 应付，处理

prosperous /ˈprɒspərəs/ *adj.*
successful or thriving, especially financially 繁荣的；兴旺的

inclusive /ɪnˈkluːsɪv/ *adj.*
including much or everything 包含的，兼收并蓄的

transform / trænsˈfɔːm/ *v.*
to completely change the appearance or character of sth./sb. 改变，使……变形；转换

sanitation /ˌsænɪˈteɪʃən/ *n.*
the process of keeping places clean and healthy 公共卫生

"More than one billion people lack access to clean water and two billion lack access to basic sanitation, and these issues require integrated engineering solutions," she said.

For example, civil and environmental engineers can build clean water infrastructure and systems. Electrical and mechanical engineers are needed to ensure these systems operate reliably.

Other engineers can develop innovative materials for low-energy water treatment, which can contribute to the sustainable goal of clean energy, she added. "Nearly one billion people, mainly in sub-Saharan Africa and South Asia, still lack access to a reliable source of electricity."

"To tackle this challenge, engineers have and should continue to find innovative, low cost solutions from renewable sources that are both accessible and environmentally friendly," she said. "Low cost, accessible solar technology in developing countries is having significant impact on the social fabric and economies of these nations."

(Words:606)

Useful Expressions

call for 要求，号召

Proper Names

Chinese Academy of Engineering 中国工程院

WFEO 世界工程师联合会

TEXT B Exercises

1. Read TEXT B and decide whether the statements are true (T) or false (F).

1) ( ) The report was issued in celebration of the 2021 World Engineering Day.

2) ( ) It is the second report of its kind in a decade.

3) ( ) World Engineering Day was created by UNESCO in 2009.

4) ( ) There is a balanced distribution of engineering science and technology resources around the world.

5) ( ) More than one billion people have access to clean water.

6) ( ) Two billion people lack access to basic sanitation.

7) ( ) Engineers should continue to find innovative, high cost solutions from renewable sources.

8) ( ) Civil and environmental engineers can build clean water infrastructure and systems.

2. Translation

Translate the sentences into English, using the words or phrases in brackets.

1) 发自该地区的各种报道表明这是一场不流血的政变。(indicate)

2) 其远期污染是不可预知或还未预知的。(predicable)

3) 这种香烟是用几种最好的烟草混合制成的。(blend)

4) 我们需要成为批判性的文本阅读者。(critical)

5) 我向我所有的学生都推荐这本书。(recommend)

6) 毕业后尼克继续深造，获得了双学士学位。(bachelor)

7) 他们两人的职业都是医生。(profession)

8) 我不是有意在名单上漏掉她的名字的，这是个疏忽。(oversight)

9) 她返回伦敦去从事她的表演事业。（pursue）

10) 他们给他房子的估价为15000元。（assess）

TEXT C

A Future City in the Making

Yang Wei shows the AR patrol glasses, Nov 4, 2020. [Photo by Chen Jiaying/chinadaily.com.cn]

Yang Wei walked into his office and put on his augmented reality **patrol** glasses every morning. Through a tiny screen inside this black **streamlined** design, the information engineer was able to immediately see the name and temperature of anyone he **encountered** in the company.

This device was of great help during the **COVID-19 pandemic** earlier this year, said Yang while connecting the glasses to his cellphone with a data cable. "We use the glasses to monitor everyone's health condition

patrol /pəˈtrəʊl/ *n.*
act of going round to check that all is secure and orderly 巡逻；巡查

streamline /ˈstriːmˌlaɪn/ *v.*
to form something into a smooth shape, so that it moves easily through the air or water 把……做成流线型

encounter /ɪnˈkaʊntə/ *v.*
to meet someone without planning to 邂逅；遇到

COVID-19 pandemic 新冠疫情

as long as the person's information is included in the company's **database**."

With such advanced technology in hand, Yang was not working in the **bustling** central business district of any large cities. As an engineer of the **5G technology** in China Construction Eighth Engineering Division in Xiong'an New Area, Hebei province, Yang spent most of his time building sites, **examining** 5G base stations and making sure the signals are stable and at work.

Three years ago, when Xiong'an New Area was first established, the region was underdeveloped with low economic growth. But it quickly **evolved** into the world's largest construction plant with a strong governmental investment, attracting workers from all over the country to devote themselves to the building of a "city of the future, a **millennium** plan".

Its importance was stressed in the Party leadership's proposals for **formulating** the 14th Five-Year Plan (2021-2025) for National Economic and Social Development and the Long-Range Objectives Through the Year 2035 on Nov 3. The Party's Central Committee demanded greater resolution in regional coordinated development and high-quality economic development of the Xiong'an New Area.

database /ˈdeɪtəˌbeɪs/ *n.*
a collection of data that is stored in a computer and that can easily be used and added to 数据库

bustling /ˈbʌslɪŋ/ *adj.*
busy and noisy 熙熙攘攘的；忙乱的

5G technology 第五代移动通信技术（fifth-generation）

examine /ɪɡˈzæmɪn/ *v.*
to look at carefully in order to learn about or from; inspect closely 检查；调查；检测

evolve /ɪˈvɒlv/ *v.*
to (cause to) develop naturally and (usual) gradually（使）逐渐形成

millennium /mɪˈlɛnɪəm/ *n.*
a period of one thousand years, especially one which begins and ends with a year ending in "000"一千年

formulate /ˈfɔːmjʊˌleɪt/ *v.*
to create (sth.) in a precise form 构想出（计划或提案）

"It is hard to make a connection between the current landscape of Xiong'an where everything just got started and is under construction, with the world's most advanced 5G technology. But the latest technology has already been **applied to** the very earliest development stage of Xiong'an," said Wang Zhicheng, a senior manager and colleague of Yang.

apply to 应用于；适用于

The company **scheduled** a roof-sealing ceremony of one of its buildings on Nov 9, to give a sense of **ritual** to a tiring life in the construction site.

schedule /ˈʃedjuːl, -ʊəl/ *v.*
to include sth. in a schedule; arrange sth. for a certain time 安排，将某事列入进度表

ritual /ˈrɪtjʊəl/ *n.*
any customary observance or practice 仪式；惯例；习俗

in charge of 负责；主管

Shi Zhaoyang, the project manager for steel construction, was **in charge of** hosting the ceremony. A native of Hebei, Shi chose to return to his home province after graduating with a master's degree from Chongqing University five years ago. "Working in the construction industry is dangerous and tiring. You spend most of the year working and living in the construction site, and can only go home on holidays."

Shi followed a fixed **routine** six days a week with one day off. He started his day at 8 in the morning and got off work at 6 in the afternoon if he didn't have an extra shift. His main duty was to make sure that every project was properly finished by time with a good quality.

routine /ruːˈtiːn/ *n.*
the usual series of things that you do at a particular time 惯例；常规

The 32-year-old engineer thought of himself as one of the luckiest people in the world this year. "My daughter was born on May 12, and I make video calls every night with her," said Shi, who **subconsciously** crossed his fingers and felt the gold wedding ring on his left ring finger.

subconsciously /ˌsʌbˈkɒnʃəslɪ/ *adv.*
from the subconscious mind 潜意识地

Living in Xiong'an, just 2 hours' drive from his home Tangshan, Shi couldn't go back home often, as he was in charge of over 130 workers and needed to ensure their safety and work. While he wasn't there all the time for his family and couldn't watch his **beloved** daughter grow, Shi was **attentive** and **rigorous** with his work, and was glad to see new buildings rising and **thriving** on the horizon of Xiong'an, a future city in the making.

beloved /bɪˈlʌvɪd/ *adj.*
person, thing, or place is one that you feel great affection for 深爱的；挚爱的

attentive /əˈtentɪv/ *adj.*
giving attention (to sb./sth.); alert and watchful 注意的；留心的

rigorous /ˈrɪɡərəs/ *adj.*
careful, thorough, and exact 一丝不苟的；缜密的

thriving [ˈθraɪvɪŋ] *adj.*
having or showing vigorous vegetal or animal life 欣欣向荣的，兴旺发达的

Unlike Shi, the 25-year-old engineer Yang Wenjun suffered from a long separation with his girlfriend, who was in his hometown in Zhejiang province. Coming to Xiong'an only three months ago, Yang felt that "the climate in Northern China is a bit challenging to me."

Newly graduated this year, Yang was still trying to **adapt to** a new life outside of campus and to a new working and living environment. While he may be **introverted**, he had a smile on his face when speaking of his girlfriend. "Actually I just came back from holiday days ago," said Shi, who spoke with a slight southern accent. "I can go home once every three months, and the company would cover my travel expenses."

Chen Jiaying is graduate student majoring in international journalism and communication at Tsinghua University, and is on a field survey tour with her classmates in Tianjin and North China's Hebei province in early November.

Chen Jiaying contributed to this story.

(Words: 759)

adapt to 适应

introverted /ˈɪntrəvɜːtɪd/ *adj*. someone who is quiet and shy and does not enjoy being with other people 内向的

TEXT C Exercise

Translate the following paragraph into Chinese.

Civil engineering is considered as the first discipline of the various branches of engineering after military engineering, and includes the designing, planning, construction, and maintenance of the infrastructure. The works include roads, bridges, buildings, dams, canals, water supply and numerous other facilities that affect the life of human beings. Civil engineering is intimately associated with the private and public sectors, including the individual homeowners and international enterprises. It is one of the oldest engineering professions, and ancient engineering achievements due to civil engineering include the pyramids of Egypt and road systems developed by the Romans.

____

____

____

____

____

____

Part III TASKS

1. Viewing and Discussing

You will watch a video clip concerning civil engineers and architects. When you finish, discuss the following questions with your partner(s). You may adopt some of the tips under each question but not limited to them.

Task Video

Question 1

What are the similarities and differences between a civil engineer and an architect?

Tips: plan and design structures, safe structure, aesthetics of the structure…

Question 2

What are the specific responsibilities of a civil engineer? What about an architect?

Tips: analyze and evaluate the structure, integrity, demanding, fulfilling, challenging, competitive...

Question 3

Are civil engineers and architects against each other? What kind of relationship should they have?

Tips: collaborative, overlap…

2. Debate

Have a debate on "which one is better, to be a civil engineer or an architect" based on your understanding of the two occupations. Try to use the expressions below in the debate.

in my opinion

personally I think

I would be glad to hear your opinion of

As you said

If I understood you correctly, you said that

Well, it depends.

I disagree with you entirely.

I agree completely.

I'm afraid I disagree.

What's your view on the matter?

练习答案

Unit 1

Part I （略）

Part II TEXT A Exercises

1. Content Questions

Keys:

1) A 2) M 3) K 4) B 5) N 6) H 7) C 8) D

2. Vocabulary

A. keys

1) illustrated 2) intelligent 3) installed 4) resistant

5) decreasing 6) reliance 7) innovation 8) ancient

9) originated 10) Establishment

B. keys

1) consist of 2) concerned with 3) as well as 4) reliance on 5) refer to

3. Translation

Key:

For a long time, the construction engineering industry was at the lower level of social cognition. To the general public, the construction engineering industry was considered labor-intensive yet low-efficiency. Construction staff has to work and live in arduous working conditions, suffer from an unstable living environment and through remote areas, far from home and family.

TEXT B Exercises

1. True (T) or False (F)

Keys:

1) F 2)T 3) T 4) F 5) T 6) F 7) F 8) T

2. Translation

Keys:

1) Cocaine addicts get specialized support from knowledgeable staff)
2) The new cars will incorporate a number of major improvements)
3) Physical Geography is a subdiscipline of Geography.
4) I think it'll depend on what type of work you do in the excavation, but I imagine we can arrange something.
5) It is an ingenious ways of saving energy
6) The whale, like the dolphin, has become a symbol of the marvels of creation.
7) This technique is particularly successful where problems occur as the result of repetitive movements.
8) There are obvious distinctions between the two wine-making areas.
9) It was an unprecedented demonstration of people power by the citizens of Moscow.
10) It is difficult to isolate the factors that led to the next development—the emergence of urban settlements.

TEXT C　Exercise

Translation

Key:

年轻技术工人的缺乏对技术工人的培养、培训造成困难。在国家层面，建议大力发展培训机构和职业技术学院，培养工匠。应该给予有技能的人员奖励。鼓励企业、行业协会等组织定期举办技能竞赛，选拔、培训和鼓励优秀员工脱颖而出，感受技艺带来的荣耀。要形成一种协作学习的氛围，促进进步和培养专业精神。

Unit 2

Part I　（略）

Part II　TEXT A　Exercises

1. Content Questions

Keys:

1) D　2) A　3) I　4) L　5) J　6) D　7) B　8) C

2. Vocabulary

A. Keys:

1) exceed	2) fireproof	3) toxic	4) chronic
5) debris	6) mechanized	7) irritation	8) culprit
9) demolition	10) asphalt		

B. Keys:

1) separate out	2) contribute to	3) add to	4) torn down
5) are, referred to as			

3. Translation

Key:

Concrete is part of the urban landscape, just as trees are part of the forest. It's so ubiquitous that we rarely pay attention to it. However, under the monotonous gray appearance, there is a complex world. Concrete is one of the most common and widely used building materials on earth. It's strong, durable, fireproof, easy to use, and can be made in any shape or size — from unimaginably large buildings to humble stepping stones.

TEXT B Exercises

1. True (T) or False (F)

Keys:

1) F 2) F 3) T 4) F 5) T 6) F 7) F 8) T

2. Translation

Keys:

1) Most of our business is done at off-site meetings.
2) The bookshelf is easy to assemble.
3) The monument was erected in honor of the soldiers who sacrificed their lives.
4) This liquid will corrode iron.
5) Careful maintenance can extend the life of cars.
6) Technological advances improve the durability of products.
7) Public figures have to withstand more pressure than ordinary people.

8) He is a versatile architect.

9) Diligence comes first when it comes to success.

10) The job is great in terms of salary.

TEXT C　Exercise

Translation

Key:

中国古代哲学认为木头能带来好运。由于风水中运用的五行理论，木材仍然是最受欢迎的建筑材料，即使是在采石和砌砖技术得以发展之后。这一理论决定了春秋时期（公元前770—476年）以来生活的许多方面。木是代表春天和生命的元素，所以对于建筑来说，它具有最好的吉祥内涵。因此，信奉风水的人认为必须用木头建造房屋等。

Unit 3

Part I　（略）

Part II　TEXT A　Exercises

1. Content Questions

Keys:

1) K　2) C　3) E　4) A　5) L　6) J　7) H　8) B

2. Vocabulary

A. Keys:

1) expertise	2) mediation	3) coordinate	4) track
5) proceed	6) assign	7) execution	8) modify
9) shift	10) lingering		

B. Keys:

1) at hand	2) familiarize, with	3) plays a major role
4) in line with	5) take shape	

3. Translation

Key:

According to a report, a construction manager has up to 120 different responsibilities during the execution of a building project. In simple words, construction managers are the ones who are responsible for the project to proceed according to the existing plan. The primary mission for construction managers is to manage their project in a way that will ensure its completion on the agreed budget and time. Furthermore, they should make sure that the whole project is complying with the set building plans, codes and other regulations.

TEXT B Exercises

1. True (T) or False (F)

Keys:

1) T 2) F 3) T 4) F 5) T 6) T 7) F 8) F

2. Translation

Keys:

1) When choosing what products to buy and which brands to buy from, more consumers are looking into sustainability.
2) The contract defines the apportionment of risks between employer and contractor.
3) These vegetables are sealed in plastic bags.
4) No one can operate such a sophisticated machine.
5) Our municipality is divided into ten districts.
6) Installation of the new equipment will take several days.
7) Prefabricating construction accessories can save time.
8) Sometimes a simple change can make all the difference.
9) Free time is at a premium for working parents.
10) Developed countries, in particular, should bear the responsibility for environmental problems.

TEXT C Exercise

Translation

Key:

从设计到挖掘，从建立互联网覆盖到开设 24 小时智能便利店，来自不同行业的专业人士共同利用他们的工业和技术专长完成了这项艰巨的紧迫任务，使医院的建设成为可能。它充分展示了中国强大的制造、建设、设计和组织能力以及专有技术。

Unit 4

Part I （略）

Part II TEXT A Exercises

1. Content Questions

Keys:

1) B 2) Q 3) R 4) Q 5) I 6) C 7) N 8) M

2. Vocabulary

A. Keys:

1) innovation 2) access 3) commitment 4) renewable

5) confirm 6) annual 7) ambition 8) enthusiasm

9) represent 10) approach

B. Keys:

1) cater to 2) take advantage of 3) scratch the surface

4) live up to 5) account for

3. Translation

Key:

This research informed the design and construction of IKON, Kingspan's global innovation center. Opened in 2019 amid the lakes and rivers of County Cavan, north of Dublin, IKON represents an investment of 10 million euros in Kingspan's commitment to a greener future. It functions as a "living experiment," providing engineers with the perfect real-world setting to measure the energy-saving properties of new materials.

TEXT B Exercises

1. True (T) or False (F)

Keys:

1) T 2) T 3) T 4) F 5) F 6) T 7) T 8) T

2. Translation

Keys:

1) It was a collective decision.
2) Her paintings capture the subtle hues of the countryside in autumn.
3) Nevertheless, this does not mean that modernization and new building should be discouraged in order to renovate and protect heritage buildings.
4) The landscape is broken only by a string of villages.
5) The hunt is on for a suitable candidate.
6) Patients are taught how to modify their diet.
7) One will never be able to understand why these political issues can acquire such force.
8) The building is part of our national heritage.
9) He is a great talent.
10) Money is not essential to happiness.

TEXT C　Exercise

Translation

Key:

今天，对环境的关注平衡了对发展的热情。曾经，建筑行业的领导者们瞄准了太空针和像飞梭一样的瞭望塔，现在他们把重点放在创新上，使建筑更环保，更能满足居住者的需求。也可以说，小处着手，方可成就大事。

Unit 5

Part I　（略）

Part II　TEXT A　Exercises

1. Content Questions

Keys:

1) C　2) A　3) E　4) I　5)J　6)K　7) G　8) D

2. Vocabulary

A. Keys:

1) alternatives　2) competent　3) alter　4) ignored

5) assessment　6) collapsed　7) slipped　8) authorization

9) defective　10) Induction

B. Keys:

1) carry out　2) at risk　3) in place

4) pay attention to　5) compare to

3. Translation

Key:

All regions, relevant departments and state-owned enterprises must clearly understand the grim safety situation facing the construction industry. We must enhance awareness of safety redlines and comprehensively improve safety in the construction industry by ensuring that contractors live up to their responsibilities, and by strengthening risk management and law enforcement especially in areas of complex geology and flood risk.

TEXT B　Exercises

1. True (T) or False (F)

Keys:

1) F　2) T　3) F　4) T　5) F　6) T　7) T　8) T

2. Translation

Keys:

1) Several people were injured, but there were no fatalities.
2) Can you give me a definite answer by tomorrow?
3) What first attracted me to her was her sense of humor.
4) What exactly is the influence of television on children?
5) This year's results are virtually the same as last year's.
6) This decision could have serious consequences for the industry.
7) We are looking for someone who is reliable and hard-working.

8) I can give you a rough estimate of the amount of wood you will need.

9) Credit cards eliminate the need to carry a lot of cash.

10) The ending of the book was entirely predictable.

TEXT C　Exercise

Translation

Key:

人工智能还使用视频信号来跟踪摄像头在建筑物中的位置，误差不超过几厘米，这样它就能识别每一帧中物体的确切位置。该系统可以每周几次跟踪大约 15 万个物体的状态。对于每个对象，AI 可以判断它处于从尚未开始安装到完全安装的三到四种状态中的哪一种。

Unit 6

Part I　（略）

Part II　TEXT A　Exercises

1. Content Questions

Keys:

1) A　2) H　3) F　4) C　5) H　6) I　7) D　8) D

2. Vocabulary

A. Keys:

1) hazard	2) consulted	3) pursue	4) critical	5) indicate
6) obtain	7) incredible	8) facilities	9) inspiring	10) assess

B. Keys:

1) dates back　2) concerned with　3) as well as　4) for instance　5) end up

3. Translation

Key:

For the most part, jobs in civil engineering are project-based. While some civil engineers will spend their working hours at the office, others will spend a lot of time in the field. However, many jobs offer a blend of field and office work. Government jobs have predictable hours, while construction work in the private sector tends to run from 7 a.m. to 3 p.m. Working for a commercial real estate firm could take longer hours and travel. If you're in a managerial position, directing projects might require longer hours as well.

TEXT B Exercises

1. True (T) or False (F)

Keys:

1) F 2) T 3) F 4) F 5) F 6) T 7) F 8) T

2. Translation

Keys:

1) Reports from the area indicate that it was a bloodless coup.
2) Besides, the long term pollution is not predicable and unforeseen.
3) These cigarettes are a blend of the best tobaccoes.
4) We need to become critical text-readers.
5) I recommend the book to all my students.
6) After school Nick went on with further study and get a double bachelor degree.
7) They're both doctors by profession.
8) I didn't mean to leave her name off the list; it was an oversight.
9) She returned to London to pursue her acting career.
10) They assess his house at 15000 yuan.

TEXT C Exercise

Translation

Key:

土木工程被认为是军事工程之后工程各分支的第一个学科，包括对基础设施的设计、规划、建设和维护。这些工程包括道路、桥梁、建筑、水坝、运河、供水系统和许多其他影响人类生活的设施。土木工程与私营单位和公共部门密切相关，包括个人业主和国际企业。它是最古老的工程专业之一，由于土木工程而取得的古代工程成就包括埃及的金字塔和罗马人开发的道路系统。